Is Science Value Free?

T0139055

'Lacey's book must be considered a major contribution and should be of interest to all philosophers of science and others interested in the role of values in supposed rational thought.'

Stephen Mumford, *Mind*

'Lacey's arguments are readily accessible and do not require a specialist's knowledge . . . the book can easily serve as an introduction to the topical and controversial question of the role of values in scientific inquiry as well as challenge taken-for-granted positions of specialists.'

James Sauer, *Research in Philosophy and Technology*

'. . . adds richness to the terms of the debate and intriguing philosophical framework for the problems that arise . . . This work will raise important questions for anyone who has wondered, not whether science currently is value-free, but what such an ideal would be and whether the idea is defensible.'

Heather Douglas, *Philosophy of Science*

Exploring the role of values in scientific inquiry Hugh Lacey examines the nature and meaning of values and looks at challenges to the view, from postmodernists, feminists, radical ecologists, third-world advocates and religious fundamentalists, that science is value free. He also focuses on discussions of "development", especially in third world countries.

Hugh Lacey is Professor of Philosophy at Swarthmore College. He is also the co-author (with Barry Schwartz) of *Behaviorism, Science, and Human Nature*.

Philosophical Issues in Science
Edited by W.H. Newton-Smith
Balliol College, Oxford

Evil or Ill?
Lawrie Reznek

The Ethics of Science: An introduction
David B. Resnik

Philosophy of Mathematics: An introduction to a world of proofs and pictures
James Robert Brown

Theories of Consciousness: An introduction and assessment
William Seager

Psychological Knowledge: A social history and philosophy
Martin Kusch

Is Science Value Free? Values and scientific understanding
Hugh Lacey

Scientific Realism: How science tracks truth
Stathis Psillos

Is Science Value Free?

Values and scientific understanding

Hugh Lacey

Routledge
Taylor & Francis Group

LONDON AND NEW YORK

First published 1999
by Routledge
11 New Fetter Lane, London EC4P 4EE

Simultaneously published in the USA and Canada
by Routledge
29 West 35th Street, New York, NY 10001

Routledge is an imprint of the Taylor & Francis Group

Paperback edition published 2005

Printed and bound in Great Britain by
TJ International Ltd, Padstow, Cornwall

British Library Cataloguing in Publication Data
A catalogue record for this book is available from the British
Library

Library of Congress Cataloging in Publication Data
Lacey, Hugh
Is science value free?: values and scientific understanding
p. cm.
Includes bibliographical references and index.
1. Sciences–Philosophy. 2. Objectivity. 3. Values.
I. Title.
Q175. L157   1999
501–dc21   00-10016
CIP

ISBN 0–415–20820–3 (Cased)
ISBN 0–415–34903–6 (Limp)

For Maria Ines, Andrew and Daniel

Like last summer's tomato harvest and many good things, this book is a product of living in our new home.

Contents

Preface to the Paperback

Motivating *Is Science Value Free?* is the view that there are rich dialectical interactions among the questions: "How to conduct scientific research?", "How to structure society?" and "How to further human well-being?" – and thence, that science may be appraised, not only for the cognitive value of its theoretical products, but also (building on this) for its contribution to social justice and human well-being. Since the hardback edition of this book was published in 1999, several books have appeared that, although they disagree about many of the details, broadly share this viewpoint (Dupré 2001; Kitcher 2001; Longino 2002a; Santos 2003). I encourage the reader to test my arguments and conclusions against those offered in these books. Nevertheless, I continue to endorse the core argument and main lines of exposition of this book. Moreover, I am confident that it retains its relevance and provides useful resources for critical engagement with these new books. For these reasons, I have made no changes in the paperback edition.

During the past five years, however, in a series of articles I have developed several of the themes of the book. Here is a list of the main developments. In the first place, the fundamental theses of impartiality, neutrality and autonomy – that, I have argued, jointly constitute the traditional view that science is, or should be, value free (chapters 4 and 10) – have been reformulated in a more elegant way (Lacey 2002a). Secondly, I have proposed additional arguments supporting that there is a significant distinction (chapter 3) between cognitive and social (and other kinds of) values (Lacey 2002b, 2004a), thus reinforcing my defense of impartiality. Thirdly, the case for the importance of conducting research under a variety of strategies has been strengthened (chapter 10). This a version of "methodological pluralism," a central theme in the books cited above that has occasioned considerable controversy; and it is linked with my thesis (chapters 6, 8 and 9) that there often exist mutually reinforcing relations between holding certain social values and adopting particular kinds of strategies in research (Lacey 2001, 2002c, 2003a, 2003b). In this way social values may play a legitimate role in scientific practices at the moment of adoption of strategy, but they have no legitimate place among the grounds for soundly accepting theories and making sound knowledge claims. Much of my new argument has built upon a detailed case study concerning research in agricultural science (anticipated in chapter 8), in which I compare research conducted under "biotechnological strategies" with that conducted under "agroecological strategies" and the different value outlooks with which they respectively have mutually reinforcing relations. I have explored at length the way in which adopting agroecological strategies is linked with social values that are widely endorsed by movements of rural workers in impoverished regions of the world. The elaboration of these values and their presuppositions provides a concrete illustration of the account of values put forward in the book (chapter 2) and has enabled its enrichment (Lacey 2002d). Fourthly, building further on this case study, I have utilized the analysis of the role of values in scientific practices provided in the book to illuminate the structure of current controversies about the use of genetically modified organisms in agriculture, and to show how research conducted under an appropriate plurality of strategies is indispensable for making sound ethical judgments on these matters (Lacey 2000, 2002c, 2002e, 2003c, 2004b).

In my opinion the case study exemplifies that an adequate philosophy of science, one suitably attentive to the role that social values may play in scientific practices, can contribute significantly to the discussion of critical issues of social ethics. It also suggests that developing such attentiveness requires dialogical contact with a variety of movements aspiring to claim space in today's world in order to express the values they espouse. Social engagement, I suggest, may positively inform the philosophy of science. Longino (2002b) has pointed out that most philosophical discussions of interactions between science and values have focused on methodological issues of science and paid little attention to values. Perhaps this imbalance derives from inadequate accounts of the nature of values that have been taken for granted by those holding empiricist-oriented philosophies; perhaps from it not having been recognized that methodological pluralism can be compatible with accepting theories, and making claims to scientific knowledge, that accord with impartiality (chapter 10). This book is intended to contribute to redressing this imbalance. It is important that it be redressed. Only then can we seriously and explicitly address the question: How can scientific practices be institution-alized so that their products nourish human well-being, one's own and that of others, with the horizon of all human beings becoming able to live fulfilling lives (lives in which the values they hold are manifested to a high degree)? Those who hold that science is value free, that its practices reflect neutrality and autonomy as well as impartiality, are confident that this question is irrelevant to (or worse, that it may distort) scientific methodology. They may also be confident that the trajectory of the current organization of science already embodies the correct answer to it. The argument of this book serves to puncture such confidence.

NEW REFERENCES

Dupré, J. (2001) *Human Nature and the Limits of Science,* Oxford: Oxford University Press.

Kitcher, P. (2001) *Science, Truth and Democracy,* New York: Oxford University Press.

Lacey, H. (2004a) "Is there a significant distinction between cognitive and social values?" in P. Machamer & G. Wolters (eds), *Science, values and objectivity,* Pittsburgh: Pittsburgh University Press (in press).

—— (2004b) "Assessing the environmental risks of transgenic crops," in *Transformação* (in press).

—— (2003a) "A ciência e o bem-estar humana: uma nova maneira de estruturar a atividade científica," in Boaventura de Sousa Santos (ed.) *Conhecimento Prudente para uma Vida Decente: Um Discurso sobre as Ciências Revisitado,* Porto: Afrontamento.

—— (2003b) "The behavioral scientist *qua* scientist makes value judgments," in *Behavior and Philosophy* (in press).

—— (2003c) "Seeds and their sociocultural nexus," in R. Figueroa & S. Harding (eds), *Science and other cultures: issues in philosophy of science and technology,* New York: Routledge.

—— (2002a) "The ways in which the sciences are and are not value free," in P. Gardenfors, K. Kijania-Placek & J. Wolenski (eds), *In the scope of logic, methodology and philosophy of science: Volume two of the 11th international congress of logic, methodology and philosophy of science,* Cracow, August 1999, Dordrecht: Kluwer.

—— (2002b) "Where values interact with science," in S. Clough (ed.) *Feminism, social justice and analytic philosophy: siblings under the skin,* Aurora, CO: The Davies Group Publishers.

—— (2002c) "Tecnociência e os valores do *Forum Social Mundial*", in I.M. Loureiro, M.E. Cevasco & J. Corrêa Leite (eds), *O Espírito de Porto Alegre,* São Paulo: Editora Paz e Terra.

—— (2002d) "Explanatory critique and emancipatory movements," in *Journal of Critical Realism* 1: 7–31.

—— (2002e) "Assessing the value of transgenic crops," in *Ethics in Science and Technology* 8: 497–511.

—— (2001) "Incommensurability and 'Multicultural Science'," in P. Hoyningen-Huene & H. Sankey (eds), *Incommensurability and Related Matters,* Dordrecht: Kluwer.

—— (2000) "Seeds and the knowledge they embody," *Peace Review* 12: 563–569.

Longino, H. E. (2002a) *The Fate of Knowledge,* Princeton: Princeton University Press.

—— (2002b) "Science and the common good: Thoughts on Philip Kitcher's *Science, Truth and Democracy*," in *Philosophy of Science* 69: 560–568.

Santos, B. de S. (2003) Santos (ed.) *Conhecimento Prudente para uma Vida Decente: Um Discurso sobre as Ciências Revisitado,* Porto: Afrontamento.

Preface

This book aims to explicate and appraise the view that science is value free: making a contribution both to analytical philosophy of science and (more speculatively) to substantive moral reflection on the place of science in contemporary society. Regarding the latter, I have discussed how science and values interact, keeping an eye toward discussions of "development," and the place of science in it, that are taking place in many "third-world" countries.

Only in passing (and through the intermediary of interlocutors whom I identify in the text) do I interact with other philosophical perspectives that discuss the interaction of values and scientific understanding, namely: critical theory, phenomenology, post-structuralism, pragmatism and social studies of science. This reflects my personal biography, not a judgment that important insights into the issues cannot be obtained from these perspectives. I hope that the readers of the book will bring my arguments into interaction with theirs.

In order to focus on my chosen themes – scientific understanding, values, and the relations between them – I have inevitably had to short-change others, concerning which of mine have presuppositions and implications. Thus, for example, I have skipped over issues about the nature of scientific theories and about how to interpret them (realism, empiricism, constructivism); and about whether scientific knowledge should be regarded as the possession of individuals or groups of individuals as social or as belonging to an abstract domain. While my arguments are intended to be independent of where one stands on the issues I do not discuss, I have not been able to develop an idiom that is completely neutral with respect to them. Throughout the book, I have used a *realist idiom*, and, except when explicitly noted, I discuss the objectives of science in *broadly realist terms*. My intention is not so much to endorse realist interpretations of science, as to show that, even with realist interpretations, which are usually bearers of the idea of science as value free, important criticisms of it can arise. I am confident that my arguments can readily be restated within, for example, empiricist perspectives. Moreover, their force does not depend upon adopting any controversial conception of the nature of scientific theories, other than that generally the acceptance of theories is of specified domains of phenomena. Thus, I believe, my arguments can be stated (criticized and appraised) without entering into current controversies about the nature of theories and realist (and other) interpretations of science.

Acknowledgments

I have worked on the interaction of values and the sciences for many years, in the course of which I have incurred many philosophical debts. My teacher, Michael Scriven, first brought the issues to my attention. Elizabeth Anderson, John Clendinnen, Richard Eldridge, J.A. Giannotti, Geoffrey Joseph, Joseph Margolis, Braulio Muñoz, Hans Oberdiek, Richard Schuldenfrei, Barry Schwartz, Miriam Solomon, Mary Tiles and dozens of my students at Swarthmore College have all been helpful from time to time over the years. As the book has been taken shape during the past three years, Elizabeth Anderson, Marcos Barbosa de Oliveira, Eduardo Barra, Otávio Bueno, John Clendinnen, Alberto Cupani, Luiz Henrique Dutra, Richard Eldridge, Brian Ellis, Ernan McMullin, Lynn Hankinson Nelson, Graham Nerlich and Howard Sankey have made useful comments and criticisms. I am especially grateful to Elizabeth Anderson, Marcos Barbosa de Oliveira, Lynn Hankinson Nelson, Graham Nerlich and Howard Sankey for correspondence about early versions of sections of the book, and to Otávio Bueno who crafted the title from a long-winded, provisional one. Miranda Fricker, as reader for Routledge, made a number of helpful suggestions.

Jackie Robinson provided assistance in all sorts of helpful ways. So, too, did the editorial staff at Routledge (Anna Gerber, Lisa Carden, Ceri Prenter). It has been a pleasure to work with them all.

During 1996, several universities provided me with extended hospitality and opportunities to test my developing ideas: the Department of History and Philosophy of Science at The University of Melbourne, Australia (especially Rod Home and Howard Sankey); the Philosophy Department at Universidade de São Paulo, Brazil, (especially Pablo Mariconda); and the Philosophy Department at Universidade Federal de Santa Catarina, Florianópolis, Brazil.

In several chapters, I have drawn upon previously published work which has been in most cases significantly rewritten and developed. Parts from

Lacey and Schwartz (1996) "The formation and transformation of values," in W. O'Donohue and R. Kitchener (eds) (1996) *The Philosophy of Psychology*, London: Sage, are reprinted in Ch. 2 by permission of Sage Publications Ltd; from Lacey (1986) "The rationality of science," in J. Margolis, M. Krausz and R.A. Burian (eds) (1986) *Rationality, Relativism and the Human Sciences*, Dordrecht: Kluwer Academic Publishers (pp. 127–50) in Chs 5 and 6 with the kind permission of Kluwer Academic Publishers; from Lacey (1990) in Ch. 6 and Lacey (1997c) in Chs 3 and 4 with permission of the editor of *Journal for the Theory of Social Behavior*; Lacey (1997b) in Chs 3 and 5 with permission of the editor of *Principia*; Lacey (1998) in Ch. 8 with permission of the production editor of *Democracy and Nature*; and Lacey (1999a, b) in Chs 5 and 6 with permission of the editor of *Science and Education*.

I am also grateful for financial support from the Eugene M. Lang Research Professorship at Swarthmore College (1993–96), the Swarthmore College Faculty Research Fund, FAPESP (Fundação para a Amparo á Pesquisa do Estado de São Paulo, Brazil), and (for earlier developments of this and related work, especially as it pertains to psychology) the National Science Foundation (1975–6, 1979–80, 1983–4).

1   Introduction

The idea that science is value free

The idea that the sciences are value free has long played a key role in the self-understanding and the public image of modern science. Poincaré, writing early in this century, captured its core as follows:

> Ethics and science have their own domains, which touch but do not interpenetrate. The one shows us to what goal we should aspire, the other, given the goal, teaches us how to attain it. So they never conflict since they never meet. There can be no more immoral science than there can be scientific morals.
>
> (Poincaré 1920/1958: 12)

Science and values only touch; they do not interpenetrate. To deny this is often perceived as to challenge that science is the pre-eminent or exemplary rational endeavor, to demean the cognitive credentials of science and to undercut its claim to produce knowledge. Lately, however, it has been much contested from an eclectic variety of viewpoints: feminism, social constructivism, pragmatism, deep ecology, fundamentalist religions, and a number of third world and indigenous people's outlooks. Exactly what is at issue does not always emerge clearly in these contestations. The rhetoric tends to be at high volume, but the argument thin. Incommensurability seems to reign. From one viewpoint, the mounting threat of multiple irrationalities and empty voluntarism looms large; from the other, the entrenchment of ideologies.

I will attempt to sort out what is at stake in the contestation of "science is value free," an idea that incorporates several distinct views about ways in which science and values do (ought) not interpenetrate. But those who affirm it have always recognized Poincaré's distinction, and held that science and values touch in various ways with more or less significant effects. Too often the critics point only to aspects of the touch, but even when the focus is on alleged interpenetrations a further ambiguity arises.

For "science is value free" in general hardly represents a fact. Perhaps it represents an idealization of fact.[1] It also represents a value, a goal or aspiration of scientific practices and a criterion for appraising its products and their consequences. The fact and value components cannot be separated. To the extent that "science is value free" represents a fact, or an idealization of a fact, that is because "science is value free" has been held as a value; and its being held as a value is without foundation if it is not possible for it to be increasingly manifest in fact. Thus to refute "science is value free" it is not enough to display cases where it is not manifested in fact; rather, the cognitive resources of the practices of science must be assessed for their ability and likelihood to bring about its manifestations increasingly and systematically.

In this introductory chapter I will provide an overview of the various sources of the idea that science is value free, leading to the proposal that it should be regarded as constituted by three component views: *impartiality*, *neutrality* and *autonomy*. Then I will outline some of the important ways in which science and values may interact without (from the proponents' viewpoint) the idea being challenged. Finally I will preview the focus, argument and methodology of the book.

SOURCES

"Science is value free" has several sources. Its kernel is present already in the works of Galileo and Bacon. Galileo (1623/1957: 270) refers to "the facts of Nature, which remains deaf and inexorable to our wishes"; and Bacon affirms, warning us to be alert to the "Idols of the mind," the sources of error to which we are prone: "The human understanding is no dry light, but receives an infusion from the will and affections; whence proceed sciences which may be called 'sciences as one would'" (Bacon 1620/1960: Aphorism 49).

Metaphysical/Galilean

The Galilean input to the idea of science as value free is metaphysical. It leads to: " ... the discarding by scientific thought of all considerations based upon value-concepts, such as perfection, harmony, meaning and aim, and finally the utter devalorization of being, the divorce of the world of value from the world of facts" (Koyré 1957: 4).

Let me summarize it in contemporary dress. The world, "the facts of nature," the spatio-temporal totality, is fully characterizable and explicable in terms of "its underlying order" – its underlying structures, processes and laws. All objects belonging to the underlying order can be fully characterized in quantitative terms; all interactions are lawful; and the laws (not necessarily deterministic) are expressible in mathematical equations. Such objects are not construed as objects of value. *Qua* objects of the underlying order, they are part of no meaningful order, they have no natural ends, no developmental potentials, and no essential relatedness to human life and practices. Values – and objects, *qua* objects of value – are not represented as emergent from the underlying order of the world.

An object may come to acquire value through its relationship to human experience, practice, or social organization, but any role it plays there is played in virtue of its causal powers, of what it is *qua* part of the underlying order of the world, so that for explanatory purposes that it may have acquired value is irrelevant. Since human beings are part of the world, some of the historically contingent states of affairs in the world will be a consequence of human causal agency. But, the view maintains, the structures, processes and laws that make up the underlying order of the world are ontologically independent of human inquiry, perception and action; they do not vary with the theoretical commitments, outlooks, interests or values of investigators. On this view, it is a "fact" that value derives from an object's relationship to human experience, practice or social organization (that human agents generate value), and so this "fact" is explicable, in principle, in terms of underlying structure, process and law. But it does not follow from a theory that explains this "fact" that human agents themselves are objects of value.

The underlying order of the world, and its constituent entities, are simply there to be discovered – the world of pure "fact" stripped of any link with value (MacIntyre 1981: 80–1). The aim of science is to represent this world of pure "fact," the underlying order of the world, independently of any relationship it might bear contingently to human practices and experiences. Such representations are posited in theories which, in order to be faithful to the object of inquiry, must deploy only categories devoid of evaluative content or implications. Thus, they must not use categories that can be applied to things only in virtue of their being related to human experiences or practices. Concretely, simplifying a little, this means using in theories only quantitative concepts, or more generally, materialist concepts (those that designate properties of material objects *qua* material objects, not *qua* related to human experience) and, in any case, no teleological, intentional or sensory concepts.

Thus we arrive at one dimension of the idea of the "neutrality" of science: scientific theories have no value judgments among their logical implications. They cannot, it is said, for they contain no value categories. A second dimension is often taken to follow: that accepting a theory has no cognitive consequences at all concerning the values one holds. A third dimension is suggested too: that scientific theories are available to be applied so as to further projects linked with any values. After all, they represent "fact" about the world, which can – so far as science is concerned – be related to, or come to serve the interests of any values whatever. If in fact they do not serve to inform the projects motivated by particular values, that is an entirely contingent matter. Notice that this last claim rests uneasily with another that has been heralded in the modern scientific tradition, that science serves especially well the projects of material progress; and it clashes strongly with those world-views (that "progress" intends to supplant) that consider the world to be infused with meaning or value.

Epistemological and methodological/Baconian

In contrast to the Galilean, the Baconian source is epistemological and methodological. Again I summarize a contemporary version. It is through experience that we gain access to the world, which can be considered a complex repository of possibilities, of which the ones that are realized may be (increasingly) connected with our practices and our planned interventions. But the world is not generally what we would have it be. Not everything that we desire or imagine to be possible is among the world's repository of possibilities. Considerations derived from values cannot determine what is possible. We find out what is possible only in the course of engagement with the world, through successful practices, including most importantly experimental ones. A scientific theory aims to encapsulate whatever it can of this repository of possibilities of the domains of phenomena within its compass; hence, the centrality in methodology of experiment.

Sound scientific knowledge, that which we can count on for practical adoption, is rooted in replicability and agreement. Only what is observed, especially in experimental settings, and certified by replication and agreement – independently of our desires, value perspectives, cultural and institutional norms and presuppositions, expedient alliances and their interests – can properly serve as evidence for scientific posits and for choosing among scientific theories. As Hempel puts it: "The grounds on which scientific hypotheses are accepted or rejected are provided by empirical evidence, which may include observational findings as well as previously established laws and theories, but surely no value judgments" (Hempel 1965: 91). This is one of the sources of the idea of the "impartiality" of science, an idea concerning the proper grounds for accepting scientific posits or making scientific judgments.

The Baconian source of *impartiality* is often complemented by a view about the nature of scientific inference, or about how empirical data are related to theories so that they can serve as evidence for accepting theoretical posits, or choosing which theories to accept. The view is that scientific inference can be reconstructed in terms of accordance with certain formal rules (Chapter 3). The rules mediate between empirical data and theories in such a way that following them leads to unambiguous choices about which theories to accept, reject or deem as requiring further investigation, or at least to unambiguous assignments of degrees of confirmation to theoretical hypotheses. They provide, as it were, the means to transfer intersubjective acceptance from the available data to the theory. While there has been at times widespread agreement that scientific inference, and any rational inference, can be explicated in terms of formal rules, there has never been anything approaching unanimity about what the rules are, or even about whether they are deductive, abductive, statistical, inductive, or some combination of these. Bacon himself is usually interpreted as holding that they are inductive.

The general view (though not any particular account of what the rules are) became reinforced with the logical empiricists' and critical rationalists' distinction between the contexts of discovery and justification, and thence with holding the rule-governed account of scientific inference to apply only in the context of justification. Oversimplifying: a theory is properly accepted (or justified) if and only if it is related to the data in accordance with the rules. Values (and, for example, metaphysics) may play a role in the process of discovery in the course of generating and exploring the merits of the theory, but they can have nothing to do with assessments of its proper acceptance. As Carnap (1928/1967: xvii) put it, after conceding a role to "emotions, drives, dispositions and general living conditions" in the process of discovery: " … for the justification of a thesis the physicist does not cite *irrational factors* [my emphasis], but gives a purely empirical-rational justification."

The success of modern science

Both *neutrality* and *impartiality* concern the content of what is posited in scientific theories: *neutrality*, its implications and consequences; *impartiality*, the grounds for accepting it. One derives from "objectivity," representing faithfully the object of inquiry; the other from "intersubjectivity" as a condition on empirical inquiry. In practice, the two ideas tend to fuse.[2] In order that there be any scientific knowledge, the Galilean idea needs to be complemented with a methodology (or procedures that can give it empirical content); and, methodologically, since objectivity cannot be had directly, intersubjectivity seems to be the best available substitute. Conversely, Baconian methodology is deployed characteristically in testing theories that meet requirements derived from the Galilean idea, although the Baconian idea itself encompasses any inquiry that is systematic and empirical (Chapters 5 and 8).

The fusion of the Galilean and Baconian ideas underlies the manifest success of modern science. Bacon promised that utility would follow from deploying his methodology. That is not what I have directly in mind. Rather it is the manifest success of modern science in increasing "the stock of knowledge." One may identify this success primarily in terms of the discovery of objects (for example atoms, electromagnetic radiation, viruses, genes) or of the definitive entrenchment of some relatively circumscribed theories (for example the heliocentric theory of planetary motion, theories of molecular chemistry, theories of the bacterial and viral causation of diseases, theories that explain the workings of instruments). Such knowledge, of course, has been widely applied in practice: in technology, in medicine, in interpreting various phenomena of the world of daily experience; and, successful application is powerful confirming evidence in support of the knowledge. Items in the stock of knowledge have been accepted in accordance with *impartiality* and so their cognitive claims are compelling regardless of what values one holds. The sustained success of modern science, as it were, speaks unambiguously to the strength (but not to the certainty or unrevisability) of its cognitive claims.

A claim that is accepted in accordance with *impartiality* is binding regardless of the values that are held – so that the presuppositions of all practices, and the beliefs that inform all actions, should (rationally) all be made consistent with it.[3] This "binding equally" should not be confused with what I have earlier called "neutrality" with its three dimensions: "consistent with all value judgments," "no (cognitive) consequences in the realm of values," "evenhandedly applicable regardless of values held"; nor with the stronger view (Chapter 3) that all practices and actions, regardless of the values they are intended to further, should be informed by scientific knowledge to the extent possible. *Neutrality* presupposes *impartiality*; and, when the Galilean and Baconian ideas are fused – especially if the metaethical and logical views described in the next two subsections are endorsed – it may appear that *impartiality* implies *neutrality*. But, I will argue (Chapter 4) "binding equally" is not consistent with endorsing all three strands of the idea of *neutrality*; and subsequent attempts to revise *neutrality* into a coherent thesis confront numerous difficulties.

Metaethical

Components of both *neutrality* and *impartiality* have been held to gain further credibility from a widespread metaethical view: that values represent subjective phenomena, preferences or utilities so that "value judgments" are considered to be only articulations of personal preferences not open to rational appraisal. As such, value judgments lack truth value: " ... they do not express assertions" (Hempel 1965: 86). A person's making them is open to scientific investigation and explanation, but not fundamentally to critical evaluation. On this view, they cannot be among a theory's logical implications, not just on the ground that theories lack value categories, but because (lacking truth value) no proposition at all can have them among its entailments. Similarly, a value judgment, in principle, cannot cognitively affect either empirical data or scientific inferences.

Logical

Closely connected with the metaethical source is a logical view: statements of fact do not entail statements of value (Hume 1739/1968); and statements of value do not entail statements of fact (Bacon's Aphorism 49, quoted on p. 2). The metaethical view is often thought to explain the logical; but the latter may be entertained in combination with other views about the nature of values. The Galilean idea may be seen as a particular instance of the general Humean schema: "Fact does not imply value," but the argument sketched for it there does not depend on affirming the general schema. The metaphysical source is independent of the logical source, and arguments (Bhaskar 1979; Margolis 1995; Murdoch 1992; Midgley 1979; Putnam 1978, 1981, 1987, 1990; Scriven 1974) against the logical view may leave the metaphysical idea untouched. On the other hand, the Baconian schema: "Value does not imply fact," seems to me to be correct and not dependent on accepting the above metaethical view. It is, however, consistent with values having implications about the interest or relevance of facts, and the adopting of values having factual presuppositions.

Both the Humean and Baconian schemata, however, draw attention away from some other logical relations involving fact and value. From fact (especially as it is represented in scientific theory) one can infer certain matters about what is possible and impossible. And judgments of value (Chapter 2) have presuppositions about what is possible and impossible. Here, at least, is an avenue through which fact and value may logically interact with important implications for working out the idea of neutrality in detail.

Practical and institutional

The fundamental sources of the idea of science as value free are those from metaphysics (ontological primacy of underlying structure, process and law), epistemology (intersubjectivity of data, rule-bound scientific inference) and methodology (centrality of experiment), and success in producing knowledge. The currency of the metaethical and logical views has provided reinforcement, rhetorically of service, but inessential. So far I have considered the idea as being about the content and consequences of scientific theoretical products, and about the character of scientific assessment and knowledge claims.

But science should not be identified with its theories. We do not grasp enough of the character of scientific theories if we abstract them from the processes in which they are generated, tested, assessed, reproduced, transformed, interlinked with other theories, adopted in practice, transmitted and surpassed. Scientific theories are both products of and of instrumental importance to *scientific practices*, and our cognitive attitudes toward theories are shaped within these practices. Members of the *scientific community* engage in these practices, which are made intelligible in the light of a long tradition; and they are conducted within various kinds of *scientific institutions*. These institutions, in turn, depend on other institutions in society at large for the provision of their necessary material and social conditions.

We can look at scientific theories from various points of view: the appropriate cognitive attitudes to hold toward them in virtue of their relations with empirical data; as products of a practice; as produced by practitioners who have certain characteristics (including qualifications and perhaps moral qualities); and as produced within certain types of institutions which express particular values, perhaps linked with those of the institutions that provide the material and social conditions needed for research or whose interests are best served by practical applications of scientific results. Since scientific practice must be conducted within institutions, the possibility of there being constraints on its conduct and outcomes, derived from the institution's interests and values, cannot be summarily excluded. The potential for tension with "science is value free" is obvious; not an idle potential, for sometimes it turns into outright conflict. I indicated an avenue on p. 7 whereby fact and value may interact through the intermediary of what is possible and impossible. A possibility presupposed by a valuative outlook (endorsed widely in society, in institutions that materially support science or by a significant political movement) may be confirmed to be impossible in a scientific theory. Then, in the name of the values, there can be a strong motive to overrule the scientific claim. Or, more subtly, where such a conflict is incipient, the scientific community (consciously or not) may simply withhold from investigating the inconvenient possibility. My point is that it is quite intelligible that values intrude on the scientific claims that are held – whether or not this intrusion is considered rationally admissible.

The idea that science is value free regards all such intrusion as distortion, and thus to be kept out of scientific practices:

> One of the strongest, if still unwritten, rules of scientific life is the prohibition of appeals to heads of state or to the populace at large in scientific matters. Recognition of a uniquely competent professional group and acceptance of its role as the exclusive arbiter of professional achievement has further implications. The group's members, as individuals and by virtue of their shared training and experience, must be seen as the sole possessors of the rules of the game or of some equivalent basis for unequivocal judgments.
>
> (Kuhn 1970: 168)

Because there is an intelligible mechanism through which such intrusion can readily occur, counter-mechanisms need to be operative within scientific practice. Thus arises the further idea of the "autonomy" of the practices and institutions in which scientific theories are generated, entertained, tested and evaluated. In practice, according to this idea, *autonomy* is a condition for gaining *impartiality* of theoretical appraisal and *neutrality* of theoretical claims.

Autonomy tends to be a rather slippery idea – reflecting diverse and even contradictory currents – and one that is often and easily trivialized. At one level it is a *political* proposal: leave science to the scientists, but also provide them with the resources to conduct their inquiries with no strings attached. Appeals to *neutrality*, and to success in gaining knowledge and informing practical applications, are often made to support this proposal. It also presumes that the growth of scientific knowledge (and of the body of accepted theories that manifest *impartiality* and *neutrality*) will take place most effectively within practices that involve and are under the control of practitioners of the scientific community. A certain reading of the *history* of science might support this presumption. *Autonomy* also draws on the idea that science has its own internal dynamic, that science defines its own problems, asks its own questions, identifies its own research priorities, seeking to gain ever more accurate, more unified, more encompassing representations of the underlying order. The internal dynamic, it is said, responds only to the evidence and to the appropriate criteria of cognitive value. According to this view, in the long run the history of science is the unfolding of this internal dynamic, punctuated by moments of intrusion from outside values and interests which always retard the process.

The proposal for *autonomy* normally grants sole authority to the scientific community with regard not only to defining problems and appraising theories, but also to determining the qualifications required for membership in the scientific community, and deciding the content of science education. This draws on the *sociological* posits that members of the scientific community conduct their scientific practice motivated by the objectives of *impartiality* and *neutrality* or, more likely, that their activities are so structured that in the long run accord with *impartiality* and *neutrality* is virtually assured; and that scientific education adequately attunes them to accept as knowledge only that which accords with *impartiality*.

These posits are bolstered by the claim, common in the public image presented by the tradition of modern science, that the scientific community has successfully cultivated among its members in their conduct as scientists the "scientific ethos" (Merton 1957)[4], the practice of such virtues as honesty, disinterestedness, forthrightness in recognizing the contributions (and opening one's own contribution publicly to the rigorous scrutiny) of others, humility and courage to follow the evidence where it leads. Clearly this is the stuff of which myths are made.

Autonomy is not easy to render in a precise thesis, and its historical and sociological presuppositions are open to further empirical investigation. It is better regarded, I think, as a reaction affirming a value in the face of unhappiness occasioned by its perceived rejection or subordination – a reaction that provides a ready and unthreatening explanation of why sometimes science has gone astray. The meaning of *autonomy* is shaped in opposition to troubling events, symbolized by the trial of Galileo, the horrors of Lysenkoism, the bemusing stubbornness of the creationists and, among some, also by the ready willingness of scientists to engage in classified research when called to do so for the sake of national security and to keep their results secret or legally limited in their use for the sake of corporate profits.

While the idea of autonomy arises as a reaction to certain kinds of "outside interferences," hinted at in symbols rather than specified sharply, there is one kind of "outside influence" that generally is tolerated, even overtly welcomed – when the institutions which fund and support science are granted an important role in determining research agendas, the problems to be investigated and the domains of phenomena to be studied. Where this happens (and it happens commonly enough) research priorities are generally not set according to the posited internal dynamic of science, but by negotiation with the bearers of non-scientific values and interests – typically for a practical reason. This need not undermine *impartiality*, though it may (Lewontin 1993), for the role of the values and interests may be restricted to the choice of research domain and need not extend to having impact on which specific theory comes to be accepted of that domain. We will see much later on p. 251 (Chapter 10) what impact it may have on *neutrality*. *Impartiality*, however, might be threatened if there was in fact an identity of (personal and social) interest among the scientific community and the agencies of support. The myth of the scientific ethos functions to deny that there are such identities of interest. Others counter that a greater diversity (of personal and social values and interests) among the practitioners of science would make a more convincing argument; but public pressure to bring about such greater diversity tends to be opposed in the name of *autonomy*.

It is a compromise of the idea of autonomy of scientific practice to grant a role to non-scientific values and interests in choosing a domain of investigation. I do not criticize such compromises *per se*, for scientific practice may be impossible without some of them (and less than complete manifestation of a value does not mean that it is not a seriously adopted value). However, in particular cases, I do criticize the choice from the perspective of other values. In its most compelling form, *autonomy* is claimed so that the responsibility of scientists – concerning *impartiality* and *neutrality* – can be exercised. In a trivial form, which has become more common in recent years, it amounts to little more than the special plea to be free to enter into compromises with whatever agencies one sees fit, without regard to the broader social interests that may be affected by the choices.

Scientific method

The ideas about empirical data and scientific inference, often functioning in concert with the Galilean idea, may be put together under the idea that modern science has a *method*. The accepted theories of modern science are the product of following a method in which intersubjectivity and often constraints grounded in the Galilean metaphysical idea are the defining elements. The method matters; not who is following the method. I mentioned on p. 10 the related idea that the practitioners of science, insofar as they engage in scientific practice, are the bearers of the virtues of the scientific ethos. *Qua* bearers of these virtues they are interchangeable, reinforcing that who is following the method does not matter, subject to the condition that the practitioner has the relevant *competencies* (observational, experimental, mathematical, inferential, conceptual, theoretical) necessary to follow the method. "Method," as used here, pertains principally to how theories come to be properly accepted or appraised, not (except as a constraint) to how they come to be put forward and entertained in the first place; it is held to pertain to the context of justification, not that of discovery.

According to common views, the other side of method is *free creativity*, for that is what supposedly enables a theory to be put forward for consideration. (A theory is created; then it is appraised following the norms of the method.) In the "context of discovery," individuality is celebrated and no potential (conscious or unconscious) stimulus to creativity (which flourishes on analogies) including values, is ruled out a priori. Perhaps values can slip in here and play unnoticed roles. Theory appraisal is comparative: it involves choice among competing theories, but the competitors have first to be "created." Values may be hidden because a competing theory that would enable the values to become manifest may not have been "created." Intersubjective agreement, obtained through following the method, may not be enough to overcome this, especially if the agreement is among practitioners selected in a context where competence is the only explicitly recognized necessary requirement, and the assumption prevails that they embody the scientific ethos. For they, the "creative innovator," the research institution and its funders may share identical interests that, in the absence of tension derived from competing interests, may simply fall into the unproblematic and unrecognized background. Thus, it is possible that values are in play and not "noticed" because the intersubjective agreement extends to include agreements about them. Here is a hint that who the practitioners are may matter.

Perhaps *impartiality* can be regularly achieved only if there is a diversity (with respect to values and interests) of practitioners in critical interaction and some diffusion of cognitive authority. "Method" may require clashing value perspectives rather than the activities of practitioners who act individually out of the scientific ethos.[5] Scientific appraisal may be communal or social: the product of interaction rather than the sum of individual acts of following the method (Longino 1990; Solomon 1992; 1994).

PERMISSIBLE INTERACTIONS BETWEEN SCIENCE AND VALUES

The ideas of impartiality, neutrality and autonomy sum up what I think is the core of the idea that science is value free. Endorsing them is compatible with values playing many roles in connection with science, most importantly: values may play decisive roles in connection with the stances adopted toward theories prior to their acceptance; cognitive values help to articulate the idea of impartiality; and the three ideas themselves function as values that may not always be well reflected in actual scientific practice.

Theories: acceptance, application, significance

Earlier, I have used expressions like "accepting" and "choosing" a theory. The idea that science is value free concerns characteristics of the theories that we accept and ought to accept, their consequences and the practices in which they are considered and come to be accepted. I will now introduce in some detail the key notion of "accepting a theory" (which is deployed frequently throughout the text) and distinguish it from several other stances that may be taken toward theories. Values may play a variety of roles in connection with the other stances.

Accepting theories

I will stipulate a usage[6] of "to accept a theory," and distinguish it from some other important stances that may be taken toward theories (hypotheses, proposals, posits, or conjectures): provisionally entertaining them, adhering to them in research practices, endorsing their greater evidential support (compared to rival theories) and applying them in practical life. A theory (T) is accepted of a domain (D) or domains of phenomena; one "accepts T of D."

To *accept T of D* is to judge that, in the light of the available evidence, T of D is sufficiently well supported that it need not be submitted to further investigation – where, for example, it is judged that further investigation can be expected only to replicate what has already been replicated many times over and to bring minor refinements of accuracy and sharper identification of the bounds of D. It is to consider it among the items of rationally consolidated beliefs (Chapter 3) or to include it in the stock of knowledge so that *ceteris paribus* it is sufficiently well established to be applied to inform practical projects (pertaining to the phenomena of D). Acceptance is a strong stance to take toward a theory. It also always remains, in principle, open to revision that might be occasioned by new developments concerning either empirical data or theory. To *reject T of D* is to accept T′ of D, where T and T′ are held to be inconsistent.

Accepting T (of D) is a stronger stance than *endorsing that*, on balance *the available evidence points more toward T than toward rival theories* that have been entertained, for then one anticipates that the balance may well be disrupted by further research including that which may provisionally entertain novel rival theories. Acceptance is a stance adopted when relevant research has become considered as effectively completed, like (ideally) in the cases mentioned (in the preceding section) as successes of modern science. In these cases it seems reasonable to maintain, and the consensus of the scientific community confirms, that further research will – not could! – not lead to a change of judgment about T, except at the levels of refinement and meeting standards of accuracy. It involves (ideally) judging that the degree of evidential support is sufficiently high according to the highest available standards for estimating it, so that consistency with an approximation of T (of D) becomes a constraint upon any theory that has more encompassing scope than T (Joseph 1980). Accepting T (of D) is accompanied by a sort of (pragmatic) certitude, but that should not be confused with (epistemic) certainty.

Accepting a theory comes at the end of a process of research in which it has been developed from predecessors, which have been provisionally entertained and adhered to by committed investigators, and separated out from both their predecessors and other rival theories by way of numerous judgments of comparative evidential support or rational acceptability – acceptance of a theory follows (properly) after having made numerous theory choices. A theory (an early version of one that may eventually be accepted) may be *provisionally entertained* for the sake of exploring its implications, its potential to generate and solve problems, and its relationship with empirical data and with other theories. Generally this involves *endorsing the plausibility* of T: To hold that T provides conceptual and hypothetical resources sufficient to shape (reshape, contribute toward) a research agenda, and that the agenda is sufficiently promising to warrant material, financial and institutional support. In order to be developed and reshaped into an acceptable form a theory must also be *adhered to*, that is, a research agenda framed by it must be participated in and commitment must be made to its furtherance.

Applying theories

Finally, accepting T (of D) is not the same thing as applying it. "Application" concerns the role of T in the realm of daily life and practical activity. T may not be applicable because D does not include significant phenomena of daily life and experience (Chapter 7) or because it cannot be deployed significantly in practices which express one's adopted value complex. To *apply* T, I stipulate, is to apply T *to* significant phenomena of daily life and experience and/or to apply it *in* practical activity. T is applied *to* phenomena when it is used (by way of providing representations of them with its categories and principles) to provide understanding of them – so that when the relevant phenomena of daily life are included in D, acceptance of T (of D) suffices to ensure its applicability to them. T is applied *in* practical activity when it is used to inform practical (often technological) activities related to the phenomena *to* which it applies concerning such matters as the workings of things, means to ends, the attainability of ends and the consequences of realizing the possible.

My usage of "apply" is more general than the one commonly associated with the phrase "applied science", which limits "apply" to the second component, that is, essentially to *technological applications*, when scientific knowledge is deployed as an instrument that informs effectively practical innovations in daily life, and particularly the development, introduction, operation and maintenance of technological devices and practices, where the outcomes of scientific inquiry become causal factors in transforming the social "world." On this common usage, the making and exploding of an atomic bomb is referred to as an application of physical theory, but explaining how the sun is a source of light and heat in terms of its thermonuclear activity is not. As a widespread and socially significant phenomenon technological application is relatively recent, dating from only about 150 years ago (White 1968), though applications of theories to explain the workings of technological objects, and theoretical reflection on technological objects as a source of scientific ideas, date at least from the time of Galileo, as do theoretical applications to numerous other phenomena of daily life and experience.

Significance of a theory

Clearly (moral and social) values must play a role when a theory is applied. No matter how strongly a theory is taken to be accepted it is applied only if applying it accords with one's values. One may apply a theory *to* significant phenomena of daily life, but not *in* one's central practical activities. In the case of technological applications, a condition of (legitimately) applying T is the moral propriety of the intended consequences and the anticipated side effects. The highest degree of cognitive value is never sufficient to legitimate practical application, so that any move from acceptance to application in practical activity should always explicitly involve considerations of moral and social values. Accepting a theory does not imply the desirability or legitimacy of applying it practically, but only that there is no cognitive barrier to doing so. The legitimate applicability of a theory in practical activity requires support from one's adopted value complex.

I will say that T (of D) is *significant for specified values* if T is applicable *to* important phenomena of daily life and experience and/or is applicable *in* practical activities in ways that further (and do not undermine) the interests shaped from adherence to the values. A theory is more or less significant for given values; and it may be highly significant, for example, concerning applications to phenomena but not concerning applications in practical activities. Significance is a matter of degree, multifaceted and subject to historical variation, and it does not follow from acceptance (cf. Anderson 1995b).

The role of cognitive values

The idea of impartiality denies that value judgments are among the grounds for accepting and rejecting theories. But to accept or reject a theory is itself to make a judgment of *cognitive value* (worth, merit) (Scriven 1991). One interpretation of *impartiality* is that judgments of "non-cognitive" (personal, moral, social, aesthetic, etc.) value play no role in choosing theories. Another, drawing heavily from the metaethical and logical sources, wants to keep out all value judgments. It proposes to do so effectively by reducing theory appraisal to the recognition of the outcomes of rule-governed operations involving formal relations between theories and empirical data. I favor the first interpretation (Chapter 3), and I will develop it in detail (Chapters 4 and 10). Regardless of interpretation, however, to choose a theory is to grant it cognitive value, to affirm (at least) that it is a better theory than some other competitor. Such affirmations are intended to be "objective"; there is a fact of the matter about which theory has greater cognitive value, causing if not outright paradox at least perplexing tension with both the metaethical and logical views that partially ground the second interpretation (Scriven 1974, 1991; Putnam 1981).

It fits with the interpretation of *impartiality* that I favor that judgments of cognitive value can be construed as the outcomes of estimates of how well theories fare when appraised in the light of certain criteria (for example, empirical adequacy, explanatory and predictive power), that is, estimates of how well theories manifest certain *cognitive values* (Chapter 3). "Science is value free," thus, should be considered compatible with the view that cognitive value judgments play essential roles in the accepting and rejecting of theories; it thus presupposes that cognitive values can be clearly distinguished from other kinds of values.

Throughout the book I will follow the *terminological convention* that the word "values," used without qualification, will mean "personal, moral, social and other values, but not cognitive values." I beg no questions of substance in doing so; should it be established that cognitive cannot be distinguished from other kinds of values the convention will have to be dropped – so too will *impartiality*. The role granted to cognitive values of course penetrates to the very heart of scientific practices. So long as no non-cognitive values penetrate in similar ways, there is nothing that the proponents of "science is value free" need regard as threatening.

Where science and values "touch": a miscellany

To use Poincaré's evocative words there are many places where science and values may touch but not interpenetrate. I will list some of what have been considered the more important places of "touch,"[7] without further comment, simply for the sake of clarifying what is, and what is not, at stake (for its proponents) in the idea that science is value free:

- Science itself is a value (not necessarily an unsubordinated one). This affirmation comes in many versions: knowledge (truth) is a value; science informs practices that produce value; its own practice requires the exercise of rationality, a universal value (Nagel 1961), or more generally, it cultivates in its practitioners characteristics that are conducive to human flourishing or well-being (Putnam 1981, 1990); it creates beauty (Poincaré 1920/1958).
- The making of value judgments, and relations among value judgments, can be informed (and criticized) by scientific knowledge of means to ends and the attainability of ends.
- There can be scientific (psychological, sociological, historical and perhaps biological) studies of values: Of their being held, manifested and embodied in persons, institutions and cultures, and of how particular values come to be held and transformed (Lacey and Schwartz 1996).
- There can be ethical evaluation of, and restrictions on, scientific practice and applications. There are, for example, ethical issues that arise in connection with the choice of research goals, the staffing of research activities, the selection of research methods (and experimental subjects), the specification of standards of proof, the dissemination of research findings, the control of scientific information, and the credit for research accomplishments (Rescher 1965: 274). Deploying a soundly accepted theory in practical application, its specific manner of application, and judgments about its "significance" reflect ethical evaluation. That a theory is sufficiently well supported to warrant its practical application, in view both of the "side effects" of applications (Scriven 1974) and of the risks of its application should it turn out to be false, involves ethical judgments (Rudner 1953).
- Values may play numerous roles (either positive or negative) in the "context of discovery," concerning judgments made in connection with the various stances that precede acceptance of a theory; in

sensitizing researchers to the importance of certain facts; in motivating research efforts (Rescher 1965); and in assessing "scientific perform-ances," such as carrying out experiments or writing papers (Scriven 1974).

- Values may play a role in connection with the compromises reached involving *autonomy* (discussed in the preceding section), for example concerning questions raised, research supported and problems se-lected; and in making judgments about whether a certain line of research should be carried out in view of probable applications that would follow.
- Commitment to certain values may motivate scrutiny of common scientific practices for "biases," focus on particular problems and policies regarding membership of the scientific community; there may be value-based criticism of scientific practices and institutions.
- The practices of science may require that their practitioners manifest certain personal and moral values (the "scientific ethos") and reinforce the valuing of certain personal traits (for example, creativity, mathe-matical and experimental capabilities). Since it has social and material conditions, it may progress, or its rate of progress may be affected at a given time, where particular social and personal values are dominant, and what these values are may vary with the historical moment (Hull 1988: 76). It may also, for the sake of the fuller manifestation of *impartiality*, require that a variety of (social and moral) values be held among its many practitioners.
- The practitioners of science may incur special moral and social responsibilities in the light of their activities and discoveries.

Science as value free: fact, idealization, or value?

That science is value free, I repeat, does not mean that there is no interplay between science and values; only that what interplay there is leaves the three component views untouched. Thus, matters of values may illuminate all sorts of aspects of the practice, sociology, institutionalization and history of science. It is not enough to impugn that science is value free to display ways in which science and values "touch" each other.

Furthermore it is not enough to impugn "science is value free" that one or other of the components is actually not highly reflected in some aspects of scientific practice. *Neutrality*, for example, may not be highly manifested in actual fact since available scientific knowledge (given current conditions) may be significantly applicable only in support of certain values; yet it may remain open to fuller manifestation in principle.

Among its proponents, each of the components of "science is value free" is itself a value, to be expressed in scientific practices and embodied in scientific institutions, a value embedded in the objectives of science itself. There is nothing paradoxical about this. As values they are manifested to varying degrees in scientific practices and in the acts of accepting or rejecting theories. Its proponents suggest that "science is value free" is also actually reflected – at least as an idealization – in many fields of science, so that in the modern scientific tradition the value, "science is value free," has become well manifested; as the slogan goes: "How else can one explain that we got to the moon?" In some fields of science, for example, psychology, it is much more difficult to back up the suggestion that it is actually manifested even as idealization. Nevertheless, as long as it is deemed possible of fuller manifestation (in a field of inquiry), it can serve as a guiding value or regulative ideal. It is not impugned unless it is shown that the trajectory of science is not or ought not be in the direction of its fuller manifestation in the making of theory choices, that it cannot or ought not serve as a regulative ideal – perhaps by showing that cognitive and other kinds of value cannot be distinguished or by arguing that there is no way to institutionalize scientific practice that can ensure that theories are chosen only in view of considerations of cognitive value.

I am not putting "science is value free" beyond the scope of critical appraisal, but pointing to the genuine complexity and multifacetedness of the interplay of science and values. The interplay is not all at the levels of logic, method or even metaphysics. How fully "science is value free" is expressed in scientific practices can depend in part on the social conditions in which research is conducted. The question, "What conditions of research are conducive to the fuller expression of 'science is value free'?" is not trivial. It is a question that needs to be informed by empirical inquiry, historical as well as sociological and psychological, that may also point to how, for example factors, other than the data and the cognitive norms, could have (unobtrusively) played a role in certain theories becoming accepted in violation of *impartiality*. (Thus, heuristically, the sociology of inquiry is relevant to the "context of justification.") I take the components of "science is value free" to be theses about the actual (past, present and future) acceptance of theories, not logical theses about an idealized science. It is idle to affirm that they represent values of scientific practice if the social conditions for their progressive fuller realization are neither present nor plausibly available to be implemented.

PREVIEW

My principal goal is to explicate and to appraise the three ideas (*impartiality*, *neutrality* and *autonomy*) that jointly constitute the view that science is value free. Thus Chapters 4 and 10 provide the foci around which all the other themes are gathered. "Science is value free" has two faces. One looks toward the sciences themselves; it deals with how values do and do not, ought and ought not, interact with the making of theory choices and with the drawing of consequences (logical and practical) from accepted theories, and the practical and institutional conditions in which they do and can do so. The other looks toward the place of science with respect to the values we hold, and how it does and does not, can and cannot, serve the projects that express these values. Often "science is value free" has been treated as little more than a footnote to discussions of scientific inference and methodology. I think it also bears on the deep value issues and conflicts of our age. Science and values are equally at the center of my attention, and so I embed my argument in an account of the general character of values (Chapter 2), and a detailed analysis of the role that scientific practices play with respect to the predominant values of modernity (Chapter 6) and of the role they might play with respect to competing values. Concerning the latter I defend the cognitive credibility of approaches to systematic empirical inquiry that are more in tune with making a variety of forms of human flourishing sustainable (Chapters 8 and 9).

My methodology involves two key steps. First, beginning from the ideas sketched earlier in "Sources," I offer provisional theses (Chapter 4) of *impartiality*, *neutrality* and *autonomy*. They are intended to provide a plausible reconstruction of the idea of science as value free that resonates throughout the tradition of modern science. The arguments of Chapters 5 to 9 provide the material for a thorough critique of these theses. Then – the second step – by way of a series of revisions motivated by these arguments I formulate (Chapter 10) new versions of *impartiality* and *neutrality* (though not *autonomy*) which I defend.

The provisional theses are already responsive to my account of the general character of values (Chapter 2), and they are formulated on the assumption that cognitive values (rather than rule-governed accounts of any kind) provide a satisfactory account of the character of the cognitive value of theories (Chapter 3). They are theses about scientific practices (and their theoretical products) which draw from the fusion of the Galilean and Baconian ideas introduced on p. 6. In the light of proposing that the objective of science is to gain understanding of domains of phenomena – explanations of them and encapsulations of the possibilities they allow – I argue (Chapter 5) that these ideas can and ought to be separated.

Then, scientific practices which draw on the fused Galilean/Baconian ideas appear as the practices of one *approach* to science, one in principle among many, one that deploys particular *strategies* – *materialist strategies* (Chapter 4) – (oversimplifying) that *constrain* the theories that are entertained to those that may represent phenomena in terms of being generated from underlying structure, process and law; and that *select* empirical data that may bear on such theories, especially data that report the outcomes of measuring and experimental operations in abstraction from the human and social contexts of the investigation. Following the materialist strategies we have been remarkably successful in identifying the "material possibilities" of phenomena, those possibilities that can be represented in terms of the generative power of underlying structure, process and law. But they do not enable us to gain access to those possibilities open to phenomena in virtue of their relations with human beings or with the social order. Perhaps (in principle) other approaches would enable us to do so, approaches (consistent with the Baconian idea) aiming to gain systematic empirical understanding appraised in the light of the cognitive values.

If (in principle) there are alternative approaches, why are the various materialist strategies deployed almost exclusively in the practices of modern science? My answer in Chapter 6, consolidated with a reflection on Kuhn in Chapter 7, points not to Galilean metaphysics, but to mutually reinforcing interactions between adopting the materialist strategies and "the modern values of control," most importantly the value of expanding our capability to exercise control over natural objects.[8]

More generally, it is not possible to pursue the objective of science (gaining understanding) except within the confines of a particular approach, where each approach is defined by the adoption of particular strategies which interact in mutually reinforcing ways with particular (social and moral) values.[9] Are there "really" alternatives to adopting the materialist strategies? Are there "really" alternative strategies under which we can gain theories that become accepted in accordance with *impartiality*? I offer (Chapters 8 and 9) two anticipatory alternatives: one – developed among some grassroots movements in third-world countries – characterized by a dialectical interplay of traditional forms of knowledge (for example, in agriculture) and materialist investigation interacts with such values as the enhancement of local well-being, agency and community, and social and ecological balance; the other, which permits an essential role for intentional categories in the investigation of human cognitive capacities, interacts with feminist values. There "really" are alternatives, so that adoption of a strategy is justified (in part) and explained by its link with values.

Values thus pervade the practices in which scientific understanding is gained; and, actually, the modern values of control are almost all pervasive. If one rejects these values, if one subordinates the place of control to other values (for example, to those of grassroots movements or to feminist ones) it is appropriate to explore alternative strategies. Values make a difference to what one does in science, to what kinds of possibilities are grasped in the theories that are produced. Thus, how science is practiced will have impact on the conditions and possibilities of daily life and experience. The phrase "science is value free" is thus misleading, and, in practice, it functions to divert attention away from the fact that strategies (as well as theories) are chosen, and that the Galilean/Baconian approach is but one approach among (in principle) many. It is best dropped. Nevertheless, there is much in its component ideas that should be refined, retained and emphasized.

The way in which values are pervasive in scientific practice is incompatible with *autonomy*. But *impartiality* remains an important, indeed essential, value of all approaches to gaining scientific understanding. Although theories are developed under strategies whose adoption is influenced by values, theories should be accepted only in the light of considerations that involve empirical data, other accepted theories and the cognitive values (Chapter 10). The levels of (and grounds for) strategy adoption and theory acceptance need to be clearly separated, while maintaining that (in the long run) links with values cannot support a strategy in the face of its failure to generate theories that are accepted in accordance with *impartiality*. *Neutrality*, with its three components: "consistent with all value judgments," "no (cognitive) consequences in the realm of values" and "evenhandedly applicable regardless of values held" turns out to be more complicated, with the second and third components requiring very significant revisions (Chapter 10). Even then, unlike in the case of *impartiality*, my final version of *neutrality* does not articulate a value that itself would be ranked high in all value outlooks; what should be the scope of *neutrality* remains controversial.

The novelty of my account derives from the identification of a set of differentiated dialectical relations between scientific practices and values.[10] It makes possible a synthesis in which due recognition is paid to the critics of "science is value free" while retaining, with a measure of re-definition, core insights of its defenders.

2 Values

If the sciences are value free, exactly what is it that they are free from? In this chapter[1] I offer a general account of values and value judgments. This account underlies the understanding of cognitive values presupposed in the statements of *impartiality*, and has far-reaching implications for the assessment of *neutrality* (Chapters 4 and 10). It also grounds subsequent substantive discussions about particular social values and their (potential) relevance to scientific inquiry.

The word "value" has varied and complex uses. In ordinary discourse, when we refer to a personal value, we may be pointing to some or all of the following:

1   A fundamental good that one pursues consistently over an extended period of one's life; an ultimate reason for one's actions.
2   A quality (or a practice) that gives worth, goodness, meaning or a fulfilling character to the life one is leading or aspiring to lead.
3   A quality (or a practice) that is partially constitutive of one's identity as a self-evaluating, self-interpreting and a partly self-making being.
4   A fundamental criterion for one to choose what is good among possible courses of action.
5   A fundamental standard to which one holds the behavior of self and others.
6   An "object of value," an appropriate relationship with which is partially constitutive both of a worthwhile life and of one's personal identity. Objects of value can include works of art, scientific theories, technological devices, sacred objects, cultures, traditions, institutions, other people and nature itself. Appropriate relations with objects of value, depending on the particular object, include the following: production, reproduction, respect, nurturance, maintenance, preservation, worship, love, public recognition and personal possession.[2]

In practical life, beliefs and desires constitute an essential part of the explanation of human action: one performs an action because one desires a certain outcome and believes that the action will contribute toward bringing about the outcome. Desires are thus among the causes of action and, as such, they may be objects of psychological and social inquiry. In addition, desires are objects of evaluation. They can be judged by the people who hold them, and others, concerning the possibility of their realization and their worth in one's own life or in a human life in general. In particular, where desires can be represented as the having of goals, agents aim, as an ideal, to have the desires that play a causal role in their behavior included among the positively evaluated desires – those that are consonant with their values.

Explaining actions in terms of an agent's beliefs and desires always presupposes a broader context in which the action in question is related to other actions (including acts of evaluation) through developing networks of beliefs and desires, which eventually make contact with the agent's fundamental goals and desires, that is, the agent's values. Through such developed explanations the causal role of values in behavior becomes apparent. Ordinary intentional explanations of action thus presuppose that values play a causal role in behavior.

PERSONAL AND SOCIAL VALUES

We may think of personal values as dialectically both the products and the points of reference of the processes with which we reflect on and evaluate our desires.[3] Holding values, then, involves second-order desires (Taylor 1985), desires about the first-order desires that play and will play a causal role in our lives; desires that only first-order desires with certain features will mark our lives as lives that are experienced as fulfilling and worthy of a human being. Holding a personal value involves the second-order desire, which represents one of a person's fundamental goals, that one's (acted on as distinct from merely felt) first-order desires be of the kinds that lead to actions that shape or produce a life marked by a certain quality (by participation in a certain practice or by appropriate relationship with a certain object of value) that it is believed makes for a fulfilling (good, meaningful, well-lived) life, and that is partially constitutive of one's identity. The role of values as criteria of choice and standards of behavior are derivative from this core meaning. Desires are personal. The *desire component* of holding values points to the personal character of values, that one's values are tied to one's most fundamental desires and to one's deepest feelings. Holding a value also involves a *belief component*, the belief that the quality referred to is indeed linked with the experience of a fulfilled life and perhaps also the belief that a life marked by this quality does not cause or rest on conditions that cause diminished lives for others.

The modes of personal values

Understood in this way, values are *manifested in behavior* whenever they, and their associated feelings and emotions, figure in the explanatory narratives that are formed to understand the behavior of an agent. Values are *woven into a life* to the extent (more or less) that the trajectory of an agent's life displays behavior constantly, consistently and recurrently manifesting the values. A value is *expressed in a practice* where conduct within the practice is furthered by and requires behavior that manifests the value. Values can also be *present* (both felt and reflected on) *in consciousness*, and *articulated in words*, representing a partial account of who one is (or would like to be or would like others to think one is), one's aspirations for the future and what one believes about human well-being and its condition.

There will always be some measure of a gap between values-as-manifested and values-as-articulated. One comes to hold values reasonably in the light of the desire and the commitment to narrow that gap. The gap has various sources. On the one hand, one's aspirations can, and often should, properly go beyond current realities. On the other hand, the gap can be a consequence of inadequate self-understanding, limited or underdeveloped capacity for self-interpretation, the desire to appear to conform to the norms of some group and even willful self-deception.

Though values cannot be reduced to their articulation,[4] their articulation is critically important. It is not just talk *about* values, as if values had a kind of being separable from their articulations, making the articulation a verbal representation of a separate, perhaps mental, reality (as a sentence describing a material object is separate from the material object). It is part of the nature of values that they be articulated. Articulation is itself an essential mode of values – part of their formation, maintenance, transformation, deepening, clarification, recognition and definition. Moreover, the very act of articulation of values may also manifest our values, since to whom we articulate our values, how and with what depth will vary according to whether the immediate audience is composed of loved one, friends, colleagues in a movement, and so on. Such articulation is part of the practice of self-interpretation, a practice necessary for a life without self-deception. It helps to define one's aspirations. It implies not only an anticipation or prediction about the future trajectory of one's life, but also promises concerning that trajectory – that the values-as-articulated will become the values-as-woven-into one's life. To be credible as aspirations, values-as-articulated must go beyond values-as-manifested currently. To be credible as predictions and promises, the gap must not be too great, for one's future possibilities are constrained by the present realities of one's life. The credibility is enhanced when one is consciously engaged in practices that offer a well-founded possibility of realizing the aspirations. There can be tension here and ample space for self-deception. Finally, the articulation of values enables values to become objects of investigation (psychological, epistemic and valuative), of reflection, of discussion and of critical argument, and when one discovers – as a consequence of articulation – that one shares one's values with others, they can become the basis of participation in shared practices and in the construction of community, the ground for living together without violence. This articulation makes it possible to reason about values; and if one does not reason about values, one will not value reason.

Embodiment of personal values in social institutions

Personal values can also be *embodied* (more or less) *in social institutions*, and *in society* as a whole. An institution embodies a value to a high degree when its normal functioning offers roles into which the value is woven, encouraging behavior that manifests it and practices that express it, reinforcing its articulation and providing the conditions for its being further woven into its members' lives. In this sense, elite universities embody to a high degree the value of intellectual cultivation and distinctiveness, but not that of solidarity with the poor; and capitalist economic institutions embody to a high degree various egoist values, but not sharing. A social order embodies a value to a high degree if it provides conditions that support institutions that embody the value, and especially if its maintenance and normal functioning depend on such institutions.

The values that can be woven into a person's life are constrained significantly by the values that are embodied to a high degree in the society in which the person lives. That is partly because articulation is an essential mode of values, and what can be articulated is a function of the linguistic resources available in one's society, which will reflect to some degree the conceptions of well-being that are dominant and reinforced in the society. This language may not readily permit the expression that one's own experience of well-being (or diminishment) does not fit well with the reigning accounts of what constitutes well-being. For example, in a society that highly embodies egoist values, in which persons are respected and recognized in virtue of their possessions, the language is not readily available for one to articulate the experience (if one has it) that manifesting such values does not produce a sense of well-being. Thence the pull is to submit one's experience to the reigning accounts of well-being, if only for the sake of being recognized and respected. In this way, values are partially constituted by the available discourse of value, and part of the reality of holding values is essentially linking one's life to the community (and its traditions) which is the source of the language of one's values.

The constraint due to the social embodiment of values exists also because people live their lives in interaction with others. Most actions are also interactions, as mentioned earlier, so that one's values will include the fundamental relations one desires to establish with others. Typically our interactions with others are mediated by social institutions: family, school, church, political and economic institutions, clubs, etc. – so that we interact in accordance with our institutional roles and with relations structured by institutions. To a considerable extent, one cannot manifest one's personal values without participating in institutions that permit their manifestation. Not every institution encourages or even permits the manifestation of one's personal values, so that whether a personal value can be woven into a life depends considerably upon the availability of institutions in which it is embodied to some degree.

I have discussed personal values as articulated in words, as present in consciousness, as manifested in action, as expressed in practices, as woven into lives and as embodied by social institutions. Values have no reality apart from these modes. In particular, personal values cannot be reduced to mental representations or simple conscious phenomena. Ontologically, values reside in the interplay of these six modes, which together are constitutive of values; and so necessarily values are developing and not simply given. They may be shared, in virtue of their being expressed in practices, articulated, and embodied in institutions – and to a considerable extent they must be. At the same time, in virtue of their manifestation in action and their being woven into individual lives, their character retains a personal element.

Kinds of values

There are various kinds of values. A value is held by an agent or agents. When an agent (X) holds a value (v), the fundamental expression is: "X values that ø be characterized by v." The different kinds of values correspond to different instantiations of ø: for example, when ø = myself, we have my personal values; ø = persons in general or relations and interactions between persons, moral values; ø = an institution, institutional values; ø = society, social values; ø = works of art, aesthetic values; and

(Chapter 3) ø = scientific theories or systematic bodies of beliefs, cognitive values.[5]

Values have both desire (want, goal) and belief dimensions: the desire that ø be (becomes) characterized by v; the belief that being characterized by v is partly constitutive of a "good" ("worthily desired") ø.

Regarding the belief dimension, depending on the ø in question, different considerations come into play. When ø = human person, the beliefs involved concern the characteristics of a fulfilled, flourishing, meaningful or well-lived human life; and they concern the relations among persons that foster (and partly constitute) fulfilled lives and that do not rest on conditions that produce diminished lives. When ø = myself, the values include characteristics that are partly constitutive of my personal (individual) identity (as discussed on p. 24). When ø = society, the values involve characteristics of social structures and organization that contribute to human well-being. No matter what ø may be, contribution to human well-being is always the "bottom line" of value discourse.

Social values and their modes

A social order is marked by the personal values that are predominantly embodied by it, and also by the social values that are woven into it. As in the case with personal (and all kinds of) values, social values involve the interplay of several modes. They are *manifested* in the programs, laws and policies of a society, and *expressed* in the practices the conditions of which it provides and reinforces. These are the values that become *articulated* in histories of the society's tradition, in explaining the kinds of institutions it has fostered, and in the rhetoric of its leadership. Again, there is always some gap between manifestation and articulation, the handling of which partially defines positions on the political spectrum. Social values are *woven into* a society to the extent that they are manifested constantly and consistently, and the gap is quite narrow. For example, liberty, the primacy of property rights and, to a much lesser extent, equality are social values highly woven into US society.

Articulation of values has a special significance in the case of social values, since typically there is contestation about social values among various members and groups in the society. Different groups within society will perceive and interpret the gap between values-as-articulated and values-as-manifested quite differently, and much of modern political discourse centers on the various competing assessments of the significance of this gap.

There is a close link between the social values woven into a society and the personal values a society embodies, and also between the values that are articulated by the dominant institutions of a society (ideology) and the personal values that become articulated throughout the society. This link need not be formal and may only become apparent as the social order unfolds concretely over time. Thus, for example, liberty (negative liberty) and the primacy of property rights, as woven into the concrete economic and legal institutions of the US, foster the embodiment of individualistic, egoistic and competitive personal values. Indeed, the embodiment of such personal values may itself be construed as a social value highly woven into the society. Under conditions in which the link between social and personal values is especially tight, the personal values people hold may come to seem natural and inevitable – so much so that they cease to be deliberated about as values and become construed as facts of human nature.

A social value can also be *personalized* when a person's acts directed toward the maintenance, modification or transformation of the social order are guided by the personal desire for a society into which this social value is woven. For example, where individualist personal values predominate, the social values of tolerance, of relations mediated by contract and of justice as fairness under the law tend to be widely personalized. This is presumably because it is believed that, given the concrete economic and legal institutions into which they are woven, these social values are among the conditions that shape a society that embodies the desired personal values. The stability of a society depends on the widespread personalizing of its predominant social values.

Furthermore, if a person's aspirations are impeded because of the prevailing predominant social values, then it makes sense to personalize other social values and to engage in political action in order to produce social forms in which they are manifested. Thus (Chapter 8; Lacey 1997c) if one aspires to express the value of solidarity with the poor, one will seek social change that would produce a social order into which positive freedom (the availability of conditions in which all have the possibility to live significant lives of their own choosing) and the primacy of economic and social rights are woven, and so personalize these social values. In this way, social values – either those predominantly manifested or those aspired to – are included among personal values.

There are, of course, differences and disagreements in the realm of values, the complexities of which cannot be cut through here. Their existence, however, highlights the question of which values one is to hold and what the relevance is of public discussion to settling the matter. Public discussion cannot be expected to result in a consensus about which values one is to hold. Indeed, a certain difference in the values people hold may be essential within the texture of an environment that can sustain human freedom. Without some diversity and tension among values, people could easily come to view the values they currently hold as the only possible values – a result that would seriously erode the scope of human aspiration and the possibilities for human development. But what public discussion can lead to is well-grounded knowledge about what are the social conditions needed for holding particular values. This is significant for it provides a causal understanding of the formation and holding of values that enable arguments to be made for the modification of existing social institutions and structures in some directions rather than others. In the light of this, I will explore a number of issues relevant to explaining the values that people do come to hold.

UNDERSTANDING THE SOURCES OF PERSONAL VALUES

Human discourse is not merely "factual"; it is not limited to providing descriptions and explanations of the way things are or have been. It is also future oriented and so contains valuative aspects. While explanation relates present states of affairs to past ones (according to causal laws or within an explanatory narrative), evaluation tends to relate the present to desired future possibilities or to their anticipated realizations.[6] Evaluation serves, in part, to set the course of our lives in the light of the constraints of actual realities. The future is neither determined by the present nor is it the product of voluntarist action unconstrained by the present. Rather, it takes shape in part as present realities are modified, and sometimes transformed, through intentional action. Our beliefs and desires play a causal role in shaping the future, but under powerful constraints that are not themselves subject to modification simply in the light of our present beliefs and desires. Values are intelligible only within this context of constraint.

Morally salient phenomena of lived experience

The context of constraint also enables the reasonably precise definition of a number of phenomena of frequent and repeated salience in everyone's lived experience. To a large degree, I hypothesize, people come to hold the values they hold in the course of responding to these phenomena, for there are a limited range of possible responses, each providing coherence to a *complex* (ensemble, set, cluster, scheme, outlook or perspective) *of values* that are woven into a person's life. These phenomena, of which I will list four, concern various gaps between aspiration and realization.

Gap between intention and effective action

The first phenomenon is related to the gap, already referred to, between the manifestation and articulation of values. It is the gap between intention and effective action, between desire and the outcomes of action. Frequently our actions do not lead to what we intend and our desires are not fulfilled through the actions they engender. Our efforts (individually or socially) to better the world do not always succeed in bettering it or producing a situation that is experienced as more fulfilling; sometimes the efforts just fail, other times they produce unintended (and undesired) effects. This gap reveals limits to our expressive capacity, our power to shape our own lives, our self-understanding, our grasp of what we can expect from others and our understanding of the social and material conditions of our lives. For example, those who propose that a life consisting of a sequence of actions based on spontaneous desires will bring a source of happiness and contentment often find that it brings instead a sense of degradation, emptiness, self-contempt and shame. Those who wish to do just "what they feel like" are often bewildered by their ineffectualness and they often discover that they fail to develop the capacities they need later on to realize desires that then take on importance for them. While first-order desires may, and often do, predate second-order desires (values) in a person's life, the continual coherence of first-order desires depends on one's developing (more or less articulately) second-order ones. There may be social or psychological conditions in which some people are unable to develop constant second-order desires and so are unable to hold values (even to hold one's self to be an object of value). Under such conditions, we might expect to find profound psychological pathology, and even little appreciation of the value of life, with consequent recourse to indifference to others or to gratuitous violence.

Gap between what we experience and what we sense can be

The second phenomenon is another gap: between what we experience to be the case and what we sense *can* be the case. We experience and observe suffering of various kinds and we sense that some of it can be mitigated, that there need not be so much suffering, that things can be "better," that the salience of suffering can be reduced and that more fulfilling possibilities can be realized. The experience of suffering, as it were, provides impetus to rank, in some sort of moral order, the possibilities that may be realized in the future. It attunes us to a sense of what well-being might be and that sense might be heightened by the observation of lives (and interaction with them) that seem to realize more fulfilling possibilities. Our experience is infused through and through with moral content. We do not experience the world merely as a sequence of facts, but as amenable to change caused by actions informed by prior deliberation. We experience it as fulfilling or lacking, as generative of a sense of well-being and of suffering – sometimes one, sometimes the other; sometimes in one respect, other times in another; sometimes better, sometimes worse; sometimes improved by our actions and social projects, sometimes worsened. Thus, it is fundamental to our experience that the world does not have to be the way it is actually, that its current state has not realized all possibilities and so we can aspire to realize other possibilities – better ones.

Different values embodied in different institutions

The third phenomenon is that each of us is placed (early in life largely by force of circumstances; later to some degree by choice) in a variety of institutions, each embodying a different complex of values. Some of these embodied values can be seen as complementary, mutually contributing toward a fulfilled life. Others "contradict" one another, setting up conflicting tendencies in the person. In the extreme case, one might be living the greater part of one's life in institutions that embody values in conflict with the most central personal values that one holds, and the values articulated in those institutions may deny credibility to the values that one is personally inclined to articulate.

Gap between values-as-articulated and values-as-manifested

Finally, within each of the institutions referred to above there is often a gap between values-as-articulated and values-as-manifested. Although institutions exist for the sake of the values they embody and articulate, in order to maintain themselves they are often pushed to pursue extraneous (perhaps important) values. For example, the core values of the university (for example, the pursuit of the truth) may find themselves compromised or overridden by the values of producing professionals to serve the current predominant order, which the university finds itself emphasizing in the service of the ends of funding and recruitment, without which it could not continue to pursue its primary values (cf. the discussions of *autonomy*: Chapters 1, 4 and 10). Institutions thus simultaneously create the conditions for the manifestation of certain values and also establish limits to their manifestation. The magnitude and severity of the tension between the values that justify the existence of an institution and the values that

enable it actually to function (MacIntyre 1981) can be expected to vary
from institution to institution and from society to society. But the tension
will always be present in some degree and, as such, will underlie the gap
between aspiration and manifestation of institutional values.

These four phenomena are among those that engulf our lives. We cannot
avoid them, though they can impinge on our critical awareness more or
less sharply. They cause disequilibrium in our lives – so much so that, to a
considerable degree, we can conceive the unfolding of a life (which
manifests values as distinct from simply desires) as the narrative of a
person's attempts to strive toward a satisfying or at least a tolerable
equilibrium. In modern times, the gap between desire and the outcome of
action is especially disconcerting because it reflects limits on our personal
freedom. In order to reduce the gap and produce a measure of equilibrium, one can attempt – always without assurance of success because the
causes of the gap (including lack of knowledge and lack of access to
required material and social conditions) might be insuperable – either to
change the shape or the social conditions of one's life, or both.

Paths toward equilibrium

I hypothesize that the paths toward equilibrium can, to a first approximation, be classified into the following five kinds and that the ordered,
coherent, unified *complex* of values that a person comes to adopt reflects the
path followed.

Adjustment

One adjusts one's goals to the way things are, the path of "realism" that
accepts (more or less consciously) that there will be no fundamental change
in the predominant institutions that shape one's life, that there are no
possibilities for the immediate future outside of the current predominant
institutions, that the future is framed by those institutions. Accordingly, one
chooses to participate – taking into account, where one can, one's
opportunities, one's education, one's family, one's talents and interests, and
one's assessment of the viability of various institutions and the further
opportunities they might open up – in those current institutions to which
one has access so as to bring about the least tension and the greatest
equilibrium. One adjusts one's goals largely to what is realizable within
these institutions, leading a life into which are woven values that are
embodied in one's society. Various ways of life fall under the path of
adjustment, reflecting the variety of institutions present, class differences

and even the existence of "fringe" niches in a society. While the path of adjustment admits of variety, within it the range of acceptable values is limited by those embodied in current dominant institutions (Lacey 1997c) – and the fact that they are socially embodied becomes *de facto* a reason for holding them or at least the ground that makes them immune to criticism.

Adopting the path of adjustment can be more or less conscious. Because the values manifested within this stance are embodied in society, normally the question of their legitimation does not arise, and if it does, the reigning societal articulations of value (ideology) quickly provide answers that *prima facie* are compelling. Adopting this stance, therefore, needs little personal reflection and, indeed, critical reflection is not a highly rated value within it, at least not critical reflection upon social structures or the kind of reflection that leads to self-consciousness within dominant practices. Critical reflection may make one become aware of the disvalue (for example, oppression, discrimination, domination) that may also be embodied in the structures, and thus intensify the sense that there are fuller possibilities waiting to be realized, and create the perception of an even larger gap between the value socially manifested and that articulated.

The path of adjustment enables, for some, the experience of a measure of fulfillment reinforced by institutional articulation (ideology) that the values woven into adjusted lives are those that define a fulfilled life (Lacey 1991a). The more stable and highly developed the structures are, the more they provide space for large numbers of people to live adjusted lives and to have their "realistically" limited desires satisfied. The stability of such structures reflects the fact that within them the actual desires of many are being satisfied, and may have the consequence that these desires appear to be fundamental and universal, reflective of human nature (Schwartz 1986).

The predominant economic and political institutions of any society reinforce the path of adjustment for a significant number of people and "privilege" those who adopt it. Rarely, however, can it be adopted by everyone, for in most if not all societies such privileged lives depend upon relations of domination, where the possibilities for the dominated to gain fulfillment are severely limited.

Resignation

One comes to resign oneself to the inevitability of the social and personal conditions of one's life, that one's desires are inefficacious, that one's aspirations are empty, that where there is change it happens outside of the operation of one's will. Then, desire becomes reduced to the desire to survive or perhaps to make life barely tolerable; and life becomes a reaction to outside forces. Here we find the phenomena of fatalism, lack of self-value and internalized oppression, diminished intelligence, suppressed consciousness and conscience, and nihilism. Resignation should perhaps be considered a "degenerate" (in the mathematical sense) path – the path adopted when the causes of disequilibrium turn out effectively to be insuperable, and so where there may be little unity and coherence within the value complexes that come to be held (and where, for many, holding values becomes reduced to simply "having" them[7]). The path does admit variety: among other possibilities it can generate gratuitous (voluntaristic) violence, the deep involvement in religious practices that transfer one's aspirations beyond the world of history, dependence on alcohol and drugs, as well as countless lives following the daily grind of survival.

There is, of course, no sharp dividing line between the paths of adjustment and resignation. Interpretive methods are needed to assign a life to one or the other type, methods that need to recognize the remarkable human capacity to find or create niches in which a meaningful life is possible. Nevertheless, resignation is a dialectical counterpart of adjustment in societies structured by dominative relations (whether the structures be economic, patriarchal, racial, or other: cf. Chapter 9). In such societies, stability requires that both paths be followed; indeed, since dominant ideology serves to disguise the structuring relations or to make them immune to critique, for most people only these two paths will concretely be available for adoption. Since the path of resignation is not conceived as having structural sources, ideology explains the adopting of it in terms of personal "defects" (laziness, lack of intelligence, etc.) woven into one's life – hence the ready willingness of the "privileged" to accept that the lives of the resigned (and, no doubt, others whom it cannot distinguish from the resigned) be "managed," and even subjected to institutional violence. Such violence is not seen as reflecting structural domination, but as protecting the value that the society should be embodying and manifesting. This conception is reinforced by the perception of various (though few) individuals moving from places where the resigned abound into the ranks of the adjusted. Unaware of structural limits, it supposes that what is available to some is available to all.

Creative marginality

The resigned are also marginalized and they make little contribution even to defining the shape of the margins. The adjusted adopt for the most part the value embodied in the current predominant institutions and so they, too, do not push the limits of the margins, but live within them. Another path accepts that the fundamental structures of society will (can, ought) not be changed in the near future, so that they will continue to frame viable lives, but reject many of the values they embody regarding them as unworthy of human aspiration. For example, it may reject the consumerist, possessivist values common in our society as degenerate versions of the aspiration of freedom; it may also be aware of the dialectic between adjustment and resignation, and react with indignation and outrage to the suffering and misery that it maintains are grounded in the prevailing structures. The response of this path is to push beyond the margins, to create spaces for the (greater) manifestation of worthier values, and for lives into which these values are woven. A number of distinct versions of this path can be distinguished: individual creativity, communal service and preservation of an alternative tradition.

Individual creativity: One recognizes that, within the prevailing structures, there are possibilities for creative expression (in art, music, science, etc.). One then pursues a talent and works on it, gaining relevant skills in order to generate something new – an object of value that is recognized in one's culture as such – that is distinctive, expressive of the self and that expands the realm of value that can be embodied within the culture. While the link between individuality and novelty makes this path compelling, it is not a solitary affair; it involves participation in shared practices, which often are institutionalized. At times the path of adjustment shades into this path. At least in some domains, the value of producing and respecting objects of value is highly embodied in most societies. Indeed, it is quite common for dominant ideology to "justify" current social structures with reference to the objects of value generated within the structures, implying that the generation of such value is sufficient to legitimize the material and social conditions necessary for its generation (Lacey 1991a). One may include in this path certain feats of entrepreneurship or creative administration. And certain corrupt practices, unusual forms of the accumulation of property and conspicuous consumption, may be considered as degenerate versions. Its defining mark is that of individual creativity in pushing out the margins, and it is not uncommon for one who follows it to find it necessary to contest the values embodied in particular institutions (for example, universities, foundations, government departments, publication houses, art museums).

Communal service: One participates in or attempts to create communities manifesting values that run counter to those of the mainstream, such as service to the needy and marginalized, and, through such communities, one seeks out ways to be less dependent on the material and social conditions that sustain the marginalization of some people. It is difficult to find space for this path. Sometimes it is found within parts of religious institutions. It is also an important factor in the women's and civil rights movements. Reforms in predominant structures may be instigated from this path, as its proponents act to make the values manifest in the dominant institutions more closely approximate their authoritative articulations, opening the path of adjustment (at least) to more of the resigned. This path also, at times, generates remarkable lives that display the creative power of radical love – the "saints," whose lives exhibit rarely realized potentials of human nature, whom we all admire even when we do not aspire to emulate them. This path may be open more to those from the marginalized sectors of society than to those from elsewhere.

Preservation of an alternative tradition: One participates in an institution or movement for the sake of preserving an alternative tradition (religious, cultural, ethnic). This can involve creating new spaces, and sometimes eventually new structures, in governmental, economic, educational and other institutions. Much of the current activity subsumed by the label "multiculturalism" in educational institutions falls into this category. This sub-path also admits of conservative varieties.

The emphasis in the first three paths is on individual change or adaptation in the light of structures that are perceived effectively as given conditions of one's life. Each of them can recognize the possibility of structural reforms, and viable structures may even need to include a place for adjusted paths committed to the administration of reforms. The remaining two paths place the emphasis on fundamental structural change.

The quest for power

This path reflects the desire to gain power (political or economic) in order to adjust social structures to one's (and, no doubt, to those one thinks that others "ought" to hold) intentions, interests and values; to use power to transform institutional structures so that one's interests or perceived obligations can be satisfied. There can be corporate, military and electoral varieties of this path.

Within prevailing structures there are roles for the exercise of power. These roles fall under the category of adjustment. The present category, in contrast, involves the use of power for fundamental structural change. Power may be used to conserve what is, to produce reform or to produce revolution. The lines between conservation and reform, and reform and revolution are not always easy to define or to identify; and one who gains an office of power may, when confronted with the realities of exercising it, move away from an intended use of power in another direction.

Transformation from below

One may hold that existing structures, even under reform, cannot provide conditions in which everyone can live lives into which are woven values that are plausibly considered their own, expressive of their human selves. One may also hold that the quest for power, at best, will only bring about changes in who occupies the privileged places in dominative structures (or perhaps replace old structures of domination with new but no less dominating kinds). Holding these views, one might enter into organizations, practices and communities (modeled on Latin American "popular organizations": Chapter 8) whose objectives are: (1) to enable their members, composed largely from the marginalized groups, to manifest values that are their own, and to practice service and cooperation to this end; (2) to expand the compass of these organizations by creating new ones and cooperating with others so that more and more people become included in the process; (3) to work with sectors from mainstream institutions in a spirit of reciprocity so as together to open up more space for fulfilling options for increasing numbers of people; (4) and in so doing, to form the institutions in which values like cooperation, participation and openness to difference can be embodied; (5) eventually to constitute the institutional base of new social structures in which relations of domination would be diminished.

This path of transformation from below has similarities to the sub-path of communal service: it begins with addressing the needs of the marginalized but, rather than service and charity, it emphasizes personal empowerment, solidarity and cooperation. Because it emphasizes the dialectic of personal and social change, it does not rest upon the agency of power. It is argued that power cannot bring about the desired changes, for power cannot make people live lives into which are woven their own values. This does not imply, however, that those adopting this path may not in actual fact sometimes become allied with groups that are using violent means to gain state power. The gaining of power by such groups is not the desired change. But, where oppression and repression are intense, expedient alliances may be judged necessary to remove crucial obstacles to progress along this path of transformation from below. Ambiguity will always be present when the followers of this path, attempting to bring about the fuller embodiment of the values it represents, interact with institutions of power. Sometimes (perhaps most of the time) the conditions of the expedient alliance will bring about a lapse into the quest for power, where the values associated with power supplant the initially motivating community values. Nevertheless, where the path of transformation from below is followed authentically, only the growth of the movements in dialectical interaction with the formation of personal values can produce the desired transformation. It rests upon a step-by-step process of change, testing each step for viability as it proceeds, a process in which there is organic unity between means and ends. It does not evaluate each step in terms of whether or not it is a means to a systematically articulated social objective, for such evaluation is unresponsive to the personal/social dialectic and open to having key roles being granted to power and violence. On the contrary, each step is evaluated in terms of its bringing about a fuller embodiment of the values articulated by the movements, and what the limits of that embodiment are; thus, of whether it represents in anticipation a society that embodies the desired values adequately and provides a ground for proceeding with fuller exploratory steps out of which the concrete structures of a transformed society can emerge (Lacey 1997c).

I suggest that these five paths (as "ideal types") are the ones that are open to people when they experience the kind of disequilibrium described at the beginning of this section. They are not pure paths. Up to a point everyone shares some aspects of all the paths, but for each person a particular path eventually comes to the fore, reflecting who the person is and what are his or her most fundamental values – the value complex that is largely constitutive of his or her identity. I hypothesize further that the first two paths, adjustment and resignation, are the most common in the contemporary world, and that to adopt any of the other paths one must have substantial motivation, because pursuing these paths often introduces new forms of disequilibrium and disorientation which, however, represent attempts to discover and realize some of the human possibilities that have not yet been realized, and to develop critical and creative consciousness in all of its dimensions.

REASONS FOR ADOPTING A COMPLEX OF VALUES

According to my analysis the path that one adopts provides a unity to the complex of values that one holds. People have various reasons, which may be more or less articulated, for adopting and persisting in their respective paths. I do not suggest that one chooses one's path as a consequence of an isolatable deliberative process, as if one deliberates considering the reasons for and against each path and then adopts one. Rather one makes choices about such matters as educational objectives, friendships to cultivate, skills to obtain, jobs or careers to pursue, places to live, commitments to family, as well as choices about countless matters of consumption and possessions, etc. – choices that are made possible, enhanced and constrained by such matters as family, class and religious background. From this multiplicity and complexity of choices emerge the contours of one's adopted path, and the ordered, unified, integrated, coherent complex of values that largely constitute one's identity. The reasons for adopting one's path become apparent not antecedently, but as one attempts to create, articulate or discern unity in the values that are manifested in the various choices and commitments that one has made, and as one looks forward to a life that displays coherence, a life into which an ordered, unified, integrated, coherent *complex of values* is increasingly woven. Such a complex of values is itself subject to evaluation in the light of a number of criteria, which often become explicit when one attempts to articulate the legitimacy of one's adopted path in the face of challenges. These criteria also play an explanatory role, at least to the extent that recognizing that one's adopted complex of values fails to meet one or other of them can occasion a life change.

I have spoken of people "holding" values and of the totality of values they hold constituting a "value complex"; where one *holds* a value (v)[8] if one desires that the relevant object (ø) be characterized by v, and believes that ø's being characterized by v is partly constitutive of a "good" ø, and one is committed to narrow the gap between its manifestation and articulation. I will say that one *adopts* a value complex if one can defend the possibility, given the constraints of prevailing material and social conditions, of each value (v) in it being more fully manifested constantly and coherently in the relevant ø (self, society, etc.) and more fully embodied in society, and if one can (to one's own satisfaction) defend the belief that the relevant ø's having v is worthily desired – where the defense will, at least in part, appeal to a view of human nature, a view of what constitutes human well-being and of what lies within human potential. Adopting a value complex, thus, has "presuppositions" that render its items integrated and coherent.

Value judgments

Adopting a value complex involves making *value judgments*, of which there are two fundamental kinds: "v is a value" ("v is worthily desired of ø"); and "v_1 is subordinate to v_2". I will say – for convenience – that value judgments (as well as values) belong to a person's value complex. Value judgments should be distinguished from judgments of the form: "v is manifested in ø to such and such a degree," and in general, any judgments about "measuring" (evaluating, estimating) the degree of manifestation of values. Often judgments of the latter type (in view of my account of the paths toward equilibrium) will play important roles in the arguments that underlie sound value judgments, and being able to make them is a condition on being able to hold values at all.[9]

Making value judgments involves responsiveness to various criteria, of which I will highlight two that are particularly pertinent for subsequent discussions of *neutrality* (Chapter 4).

- *The possibility criterion*: The genuine possibility – given the actual constraints of prevailing material and social conditions – of the complex of values being woven consistently, constantly and coherently into a concrete personal life (or other relevant ø), and so of being embodied in society.
- *The human nature criterion*: The availability of an articulated view of human nature, with some empirical support, that renders intelligible the claim that holding the complex of values shapes fulfilling lives.
- In practice, these two criteria are deployed in conjunction with a number of other criteria, several of which I list without elaboration and defense.
- The formal consistency of the value complex.
- The continuity of the value complex, perhaps under considerable reinterpretation, (a) with some of the values that one has "inherited," that one shares with some others and that appear "obvious" (for example, rejection of murder) and also (b) with values that are actually woven into concrete lives (or compelling literary characters) that one recognizes as fulfilled.
- The inclusion in the value complex of those values that are constitutive of the intelligibility of valuative discourse in general (for example, respect for the participants in the discourse, elevation of dialogue over power, truthfulness).
- The universalizability of the core values of the complex – or rather, that the material and social conditions of their embodiment are compatible with all participants in the discourse being able to conduct lives into which these values (or, more generally, those expressive of their own individual identities) are woven.

I have left it open whom one considers the participants in valuative discourse to be, for that – perhaps – reflects one's own values, rather than a general criterion. If everyone (in principle) is included among the participants then, as the final two criteria are met, one gains a greater capacity to understand well the actions of other persons and social movements, even those that manifest values quite different from one's own. Gaining such a capacity is a core value of the fifth path – the path of "transformation from below" (Lacey 1997c). So, too, is gaining a clear consciousness of the conditions necessary for one's values to be woven into one's life.

Whatever one may think of the other proposed criteria, the possibility and human nature criteria are of utmost importance for evaluating the value complex that a person holds, although they do not function independently of attempts to gain "reflective equilibrium" (to borrow Rawls' term) with other criteria. The view of human nature underlies the coherence and intelligibility of the complex; the genuine possibility of its being held is necessary for its social salience. Frequently, these two criteria lie at the heart of valuative disputes functioning in concert, for example, arguments about the possibility or impossibility of social transformation often appeal to a view of human nature and the possibilities encapsulated in its articulation. Moreover, if one articulates values not embodied in the current social order, and it is not possible to generate social transformation of the desired kind, then that is a reason to reconsider one's aspirations. If social transformation is possible, then that a rival value complex is embodied in the current order ceases to be a compelling argument for one's aspirations being limited by it (Lacey 1991a, 1997c). If arguments can be mounted simultaneously that social transformation is possible and that there is a supported view of human nature that suggests more fulfilling possibilities in the proposed new order, then the value complex embodied in the current order may come into crisis and, for want of coherence and salience, the extent of its embodiment may decline.

If valuation is to guide lives and not become reduced to mere idealistic criticism, determinations of what is possible are always very important. Formal consistency of a value complex is not sufficient for such determinations. For example, community values like cooperation and sharing are formally consistent with the primacy of property rights, but arguably the material and social conditions required for the steadfast manifestation of one precludes that of the other. More than consistency is involved in assessing the possibility of coherently holding a value complex. When a value complex is already highly embodied in a society, there is no further question about its possibility. What is actual is possible. A particularly difficult question confronts us when the fifth path, the path of social transformation, is under consideration (Chapter 8). Unless the kind of social transformation proposed is possible, the path of social transformation reduces to that of communal service. In our society, this kind of social transformation is widely believed to be impossible, so that most lives can be charted on the first three paths. It is widely believed that any viable social possibilities (for the foreseeable future) will be framed by the institutions of private property, the free market and formal, electoral democracy (Lacey

1997c). Why? There are various proposed answers: (a) because those institutions are arguably better (more fulfilling, more encouraging of human freedom, more conducive to social justice), and recognized as such, than any available and most imaginable alternatives; (b) because of their inertia and ever growing momentum; (c) because of the virtual hegemony of power associated with them and the expectancy that this power will be used to maintain their hegemony; (d) because of the belief that they effectively embody human nature, which is understood to underlie individualist and egoist values. If (a) and (d) are soundly grounded, then the use of power, referred to in (c), will often be legitimate – but not otherwise. Much hinges, then, on an assessment of what human nature is and an assessment of what makes it that way.

It accords with the two criteria under discussion that beliefs about what is possible and about human nature may properly influence the personal values that one holds and desires to be embodied in society, and the social values that one attempts to personalize. Indeed, the applicability of the criteria depends upon such beliefs. As already indicated, beliefs about what is possible and about human nature are deeply intertwined. Our beliefs about the values that can be woven into lives and articulated authentically depend (in part) on our beliefs about human nature; and our beliefs about human nature draw heavily upon the values that we observe to be actually woven into lives. Broadly speaking, these beliefs are open to empirical scrutiny (and, because of this, the proposed component of *neutrality*, that established scientific theories can have no consequences for the values one holds, cannot be sustained: Chapter 4). What kind of empirical inquiry could throw light on them, and can its results accord with *impartiality*?

Answering this question is made exceptionally difficult by the fact that the possibility of genuinely holding values depends not only upon what human nature is, but also upon the values that are embodied in actual societal institutions and upon the power (authority) relations that structure these institutions. Prevailing power relations may actually prevent certain possibilities allowed by human nature from being realized, especially in those cases where their being realized rests upon social conditions that are incompatible with the prevailing conditions. Moreover, the human desires that are present at a given time might reflect not the full potential of human nature, but the possibilities reinforced in institutions. If this is so, then an empirical charting of what is actually manifested (and empirical inquiry must be based in observation of the actual) cannot result in a comprehensive account of human potential, for there may be hitherto unrealized possibilities.

What kind of empirical (psychological, sociological) inquiry can properly inform beliefs about the possibility (not necessarily high probability) or impossibility of realizing value complexes, such as those implicit in the fifth path, that have not hitherto been realized to any significant extent? What is the appropriate methodology? What are the appropriate theoretical constraints? What are the appropriate empirical phenomena to select to investigate? Answers to these questions are usually intertwined and reflect the practices (and the values they express) that the research informs (Lacey 1990, 1997c). If most research informs practices within our first three paths, then the question about alternative possibilities is not likely to arise (expect at the margins). If today these paths are framed by the institutions of private property, the free market, etc. and those institutions tend to embody egoist values, made coherent and intelligible by an individualist view of human nature (Lacey 1997c), then we would expect to find psychological and social theory today constrained by the individualist view of human nature and its evidential base would draw principally from the characteristic phenomena of these institutions, where one observes the predominance of egoist values (Lacey and Schwartz 1996). The prospects for accord with the *impartiality* of such investigations will be raised in Chapters 8 and 10.

3 Cognitive values

In the grammar of "values," the most fundamental expression has the form: "X values that ø be characterized by v," where "X" designates a person and where different kinds of values correspond to different instantiations of ø (Chapter 2). When ø is instantiated by a person's beliefs or by an accepted scientific theory, the "v" designates *cognitive values*.

THE IDEA OF COGNITIVE VALUES

Cognitive values are characteristics (criteria) of "good" (rationally acceptable, desirably held) beliefs and "good" (soundly accepted) theories. I will consider them first, in the most general way, general as desiderata of beliefs. Beliefs (about, for example, human nature and what it is possible to achieve) are presupposed in adopting a value complex (Chapter 2). They are also involved, together with desires, in the generation of action: one acts because one desires a certain outcome and believes that the action will further the realization of the outcome. One's success in adopting values and weaving them into one's life, therefore, rests in part upon holding the right kinds of beliefs; not just "significant" or "relevant" beliefs, but those that have been gained, sorted and ranked in the light of the ideal of truth.

Not all beliefs are true, of course, just as not all desires are good. But it would be contradictory to affirm simultaneously: "p is false and I believe that p." Truth does not display itself as a manifest property of the beliefs one affirms to be true, just as goodness does not display itself as a manifest property of what one affirms to be good. One judges a belief to be true in virtue of whether one believes it to possess certain properties and relations with other beliefs (however they may be elaborated in detail): to be well grounded in evidence, to follow by entailment from other true beliefs, to satisfy rational canons, to bear the appropriate inductive and deductive relations with other beliefs, to have a particular kind of causal history, to inform action that is successful consistently – in other words in virtue of an appraisal of its cognitive value. The properties and relations of beliefs that one identifies as serving this function are (to a first approximation) one's *cognitive values*. Clearly most of them may be manifested in one's beliefs to a greater or lesser degree; the greater the degree to which they are manifested, then the greater is the "rational acceptability" of the belief. Judgments of degree of rational acceptability are framed by the ideal of truth, but we have no indicator of truth other than rational acceptability.

Grammatically and logically, cognitive values have much in common with other values. Holding a cognitive value v, involves second-order propositional attitudes: a belief about beliefs, a second-order belief that being characterized by v contributes to making a (first-order) belief rationally acceptable; and a desire that *ceteris paribus* one's beliefs be characterized by v. Thus, v represents an attribute that one desires one's beliefs to possess and one believes that they ought to possess, in part for the sake of being able to live a life that expresses one's fundamental values. Then, holding a set of cognitive values implies a commitment to evaluating the beliefs that inform one's actions to the extent appropriate and possible in the light of assessments of their manifestations of the items of this set – while recognizing that there will remain inevitable ambiguities, gaps and lapses, and that the exigencies of action often curtail deliberation. The cognitive values one adopts form an essential part of one's total value complex. For some, it may be the most fundamental or highly ranked part. Putnam, for example, maintains that the cognitive values constitute part of our idea of rationality and of total human flourishing, so that aiming to weave them integrally into one's life represents a value that may (ought to) be manifested and articulated quite apart from the immediate contexts of practical action (1981: 134–6; 1990: 139–41).

Beliefs

Beliefs are propositional attitudes that, together with desires, intentions, having goals and the like, may play causal roles in generating actions. In the grammar of "belief," the fundamental expression is: "X (an agent) believes that p." A belief is always a belief of an agent or shared among agents. It is true if, and only if, its propositional content is true. Thus, the critical evaluation of the belief that p is identical with the cognitive (rational) appraisal of p.

The causal role of beliefs is represented not in law-like schemata but in "practical syllogisms," in which one's actions are represented as following (rationally) from having certain goals (desires) and beliefs (Lacey 1996; Lacey and Schwartz 1986; and especially Donagan 1989). A typical practical syllogism includes one's beliefs about the character of the situation one is in, about the means to the desired outcome in that situation, about the side effects (desired and undesired, intended and unintended) of the action, and about the possibility of realization of the desired outcome. I will say that a *belief informs an action* if it is among those of the practical syllogism that serves to explain the action and that it *informs one's activity* if it regularly informs one's actions and/or it is among the presuppositions that play a role in the articulation and justification of one's value complex. Here, I include under "belief" any of the cognitive attitudes that agents enunciate as beliefs, or that accompany desires (as represented in practical syllogisms) in the causation of their actions.

Agents' beliefs themselves have causes – more accurately, that an agent believes that p has causes – which may, or may not, include their assessments of evidence and other explicitly cognitive or rational factors. The causes of an agent's beliefs may, or may not, be reasons to hold them. They may include a wide variety of social, psychological and experiential factors; for example, one's immediate or reflected-upon experience, the testimony of "authorities," the inheritance of a tradition and efforts to make intelligible a value complex to which one is attracted.[1] These help to explain the fact that often one's actions are informed by false beliefs.

Attention to the causal role of beliefs is insufficient to evaluate them, that is, to appraise whether or not they manifest adequately the relevant cognitive values. To make this clear, I will distinguish: one *has the belief* that p if that belief informs one's actions (including verbal acts of enunciating beliefs). One *holds the belief* that p if one reflectively endorses that p and it informs one's activity; and one can defend p from criticism to one's own satisfaction by pointing to its place in a network of evidential and logical relations, articulated by one's adopted cognitive values. One *holds that the belief that p is consolidated* if one judges that p belongs to the class of rationally acceptable beliefs, those that (methodologically) require no further investigation, like theories of certain domains of phenomena (Chapter 4) which manifest all of the appropriate cognitive values to a very high degree. An agent may have but *ceteris paribus* not hold both the beliefs that p and that ¬p. Holding, rather than just having a belief, implies that one's cognitive activities of assessment of evidence and argument are among the causes of (maintenance of) the belief. That does not imply that psychological and social causes are absent or of low salience. Social causes, for example, may explain the limited range of evidence one has considered. Holding a belief requires responding to the criticisms one encounters to one's own satisfaction; holding that a belief is consolidated requires a fuller articulation of the evidential and logical network in which the belief is inserted, and follows from having actively sought out criticisms among the relevant community of inquiry and having arrived at a virtual consensus with its members concerning both the belief and the appropriate cognitive values. Beliefs that are held to be consolidated are often referred to as "knowledge."

There should be, it seems, an identity between the beliefs one has and those one holds; or at least, it seems, if one holds the belief that p then it would be irrational to act in ways that were informed by ¬p. I will call this an "informal ideal of rationality." But it is not, in all situations, part of the ideal that there be an identity between the beliefs one holds and those one holds to be consolidated, because in many of the contexts in which we must act beliefs held to be consolidated are not available, and because genuinely creative activity and activity driven by our deepest hopes and most fundamental values, and that which aims to realize hitherto unrealized possibilities (whose realizability depends causally upon committed action: Lacey 1997c) will necessarily be informed by beliefs which do not manifest *all* of the cognitive values to high degrees. Presumably, however, when one holds certain beliefs to be consolidated, and they are relevant given the context and immediate objectives of action, one's actions should be informed by them. The ideal includes *ceteris paribus* the desirability of expanding the stock of beliefs that one holds to be consolidated. It is not always rational to act informed by the best available beliefs that are held to be consolidated; our objectives should be shaped by our values without curtailing them to be informed by (though they should be consistent with) available scientific knowledge. It follows that the evaluation of beliefs, that are pertinent to planning actions, is responsive not only to the ideal of truth, but also to relevance or *significance*; that is, whether it has content appropriate to inform actions aiming at the proposed goal. But considerations of significance do not render the ideal of truth irrelevant, for alone it is insufficient to inform successful action (Chapter 9), and the judgment of insignificance does not imply that of falsity. If available scientific knowledge is not relevantly applicable for the pursuit of one's objectives, however, there is often good reason to attempt to gain well-founded, empirical knowledge that would become so applicable (Chapters 8 and 9).

Sometimes actions are partly caused by false beliefs; but if one acts based on the belief that p, and then one discovers that p is false, one says, "I believed that p was true but I was in error." One does not tend to lapse into subjectivism here, and say "p is (was) true for me." Similarly where people hold opposed beliefs which the available evidence does not adjudicate decisively, we do not relativize truth to the persons. Rather we say that they have different opinions concerning whether the proposition affirmed in the belief is true or false (even if the significance criterion explains one person holding the belief, but not another). In this respect the grammar of beliefs diverges from that of desires. That X desires that p, but Y desires that ¬p need not imply conflict or contradiction; that X believes that p, and Y believes that ¬p, does.[2] This is reflected in expecting that the cognitive values one comes to hold should make a rational claim on everyone.

Often it is the falsity of the beliefs that inform action, not the agent's belief itself, that explains some important feature of the context of action. Having false beliefs about means to desired outcomes can lead to frustrated desires; having them about side-effects can underlie destructive consequences of one's actions; and having them about the possibilities of outcomes and the presuppositions of practices can be responsible for futile efforts, for desirable projects not being entertained or supported (often entrenching conditions of widespread suffering: Chapter 8) and for assuming self-serving objectives. These familiar kinds of facts reinforce the close connection between the cognitive values and other values. We cannot understand our own and other people's actions, when they succeed and can succeed in producing their desired effects, and when they contribute and can contribute to the expression of our values, unless we have effective methods for evaluating beliefs. Without such effective methods, the commitment to a life into which one's values are woven is weakened.

Holding, not just having, beliefs is essential for being able to weave one's values into one's life. One holds only those beliefs that have been evaluated, and thus put into a coherent network of beliefs, articulated through reference to one's cognitive values. Beliefs may be held with varying degrees of confidence, often marked in ordinary discourse by using terms like "know," "probably," "sure," "hunch," "opinion," "think," "speculate," "conjecture" and "hope"; and reflected in the distinction between holding beliefs and holding beliefs to be consolidated. Held beliefs are a subset of the beliefs one has. The former generally arise from the latter or from consideration of the beliefs of others that one has been caused to be aware of. The strengthening of the beliefs one holds often happens in the course of rejecting conflicting views. Evaluating the beliefs one brings to the planning of action, therefore, ought not abstract from considerations of their causation and the causation of other candidates for beliefs that have been rejected. In particular, it is important to raise the question of whether one has considered a suitable range of candidates for belief.

Beliefs, like values, are present in various modes: they are among the causes of action; they may be articulated in words; in virtue of which they can become objects of evaluation and become held (or rejected). Articulation is a necessary condition for evaluation (whether responsive to the ideal of truth or to significance); it enables a belief to be considered in inquiry largely in abstraction from its causal roles. Rational agents aim to minimize the gap that exists between the beliefs they hold and those that inform their actions, so that they act to the extent possible only upon held beliefs. The interplay between beliefs as causal factors in action and beliefs as articulated and held is characteristic of the nature of beliefs (Lacey 1996). In order to identify the beliefs that people have, we must attend both to the beliefs that we discern (on interpretive inquiry) to be the causes of their actions, and to those they verbally enunciate and present reasons for holding – for there may be unarticulated beliefs, and a person's own enunciation is insufficient to identify his or her beliefs. A clear identification is possible only when there are no gaps between the two; and in the case of held beliefs only when also their place in a network of beliefs, linked by logical and evidential relations or articulated by their adopted cognitive values, is made clear. Of course, one may hold beliefs and properly affirm that they are well supported, even though they play no role in one's actions (other than perhaps in verbal acts). These will be beliefs gained in the course of elaborating the network of beliefs that one holds. Theory extends beyond practice and one's curiosity extends beyond the immediacies of one's experiences and actions.

Modes of cognitive values

As with values in general, and with beliefs, a person's cognitive values function in several modes.

Cognitive values as manifested in patterns of beliefs

Cognitive values are *manifested* in the networks of first-order beliefs that one holds insofar as the patterns among one's held beliefs (and the kinds of beliefs that one considers and rejects) are intelligible only in virtue of (perhaps implicit) beliefs about such matters as the links among acceptable beliefs, which beliefs constitute evidence for other beliefs, what the evidential relations are and which causal sources readily engender acceptable beliefs. For example, one manifests the cognitive value "predictive power" if one holds various beliefs, because from them reliable predictions have been made concerning other items that are believed because they were gained in the course of direct observation. Clearly this cognitive value is manifested to a greater degree as the range of vindicated predictions in one's belief network is greater, and where beliefs become held because of the relations among beliefs that enable successful predictions to be made. In this respect, predictive power is typical of many cognitive values; it may be manifested more or less. Identifying the cognitive values that are manifested in one's beliefs, like identifying the values manifested in one's behavior, can involve complicated interpretive activity.[3] To the extent that a set of cognitive values is manifested in one's beliefs to a high degree constantly, consistently, recurrently and more so with the passage of time, they are *woven into* one's life.

Cognitive values as partly constitutive of belief-gaining practices

One comes to hold beliefs in the course of engagement in "belief-gaining practices," the objective of which is the generation, selection, evaluation and consolidation of acceptable beliefs. Cognitive values are *expressed* in and *partly constitutive* of such practices; one engages in them because they lead to holding the desired kinds of beliefs, those which manifest the cognitive values.

Belief-gaining practices may involve the use of observational, inferential, conversational, argumentative, listening, scholarly, interpretive, historical, linguistic, imaginative, mathematical and other skills and procedures. They include both causal (generating candidates for belief) and evaluative dimensions. They may not always be identified by those who engage in them as belief-gaining, and generally they may not be separable in people's lives from the other practices in which they engage. They may simply be an integral part of other practices; one may engage in them either spontaneously or because they provide the relevant kind of beliefs needed to inform action within these practices. When a belief-gaining practice is explicitly identified and institutionalized (as in the sciences), engaging in it involves adopting the cognitive values that are partly constitutive of it and sharing them with other practitioners. Then, in the context of this practice, if one is authentically engaged in it, the cognitive values manifested in one's beliefs will be identical with those that are partly constitutive of the practice.

Articulation of cognitive values

One can also *articulate* one's cognitive values. Thus, they can be laid out for public scrutiny. While it is a commonplace to articulate beliefs, usually the articulation of cognitive values is an indication of engaging in philosophical reflection. Nevertheless, categories used to articulate cognitive values do play a more or less rigorous role in daily discourse when one uses, for example, "evidence," "probable," "predict," "consistent," "certain," "explain" and other terms that come into play when asked why one holds certain beliefs and when challenging those of others. There seems to be considerable overlap in the cognitive values that people have. I conjecture that there are few who do not have the following under some interpretation or other: direct origin in one's own or a trustworthy informant's perceptual experience; explanatory, anticipatory and predictive power among beliefs originating in experience; inductive derivability – all, of course, *prima facie* and to be brought into an appropriate reflective balance. Without some such shared cognitive values, under shared interpretations, communication would be impossible, and where it cannot be counted on there is defective communication (Lacey 1991a).

Disagreements about cognitive values

Often there is, alongside considerable agreement, disagreement about what should be included in the list of cognitive values, about their relative rankings, and about how adequately some of them are manifested in certain bodies of beliefs or scientific theories. Such disagreements can be compatible with reasoned dialogue and arguments about, for example, what cognitive values ought to be held and whether one ought to deploy the same set of cognitive values regardless of the particular practice or activity that one wants one's beliefs to inform.

Disagreements about cognitive values obviously will be implicated in conflicts about the beliefs that people come to hold. Since there are such disagreements, it follows that people – even after careful deliberation – can end up with conflicting beliefs. (I do not imply that agreement about cognitive values ensures agreements about beliefs.) When this happens, it is simplistic to urge people to solve their differences by more careful scrutiny of the evidence. It does not follow that the disagreements indicate that at least one of the parties' beliefs must rest upon such things as blind authority, bias, making premature judgments, or uncritical reliance on tradition; for among the interacting factors that causally account for conflicting beliefs being held are: 1) the different causal sources (social, psychological, experiential) of the beliefs; 2) their bearers holding different sets of cognitive values; 3) and related to this, their engagement in different belief-gaining practices.

The fact of such disagreements, and their persistence, does not signify that truth ought to be relativized to the believer. Putnam argues that holding different sets of cognitive values should be regarded as signifying disagreements and not simply differences, for it can lead to holding contradictory beliefs (1981, 1987, 1990). It is difficult to get away from the sense that there is a correct set of cognitive values that one ought to aspire to identify.[4] The aspiration need not presuppose that the creation of new belief-gaining practices will not lead to surpassing, or radically re-interpreting, currently available lists of cognitive values, and it is compatible with there being a dialectical interaction between holding cognitive values and historically located belief-gaining practices (Bernstein 1983; Laudan 1984). Unlike with some personal values, where holding different values represents only difference and not disagreement, since different people properly aspire to lives into which different personal values are woven, differences in the cognitive values that are held always represent disagreements. One's personal identity is not linked with having one's "personal" set of cognitive values. My living a life of integrity and authentic identity seems to require that I hold not "my" cognitive values, but the right ones. The cognitive values one holds, it seems, are right or wrong; and so, throughout this book, I often refer to *the* cognitive values without specifying who holds them.

Disagreements about which cognitive values to hold are very common, even when discussing the sciences, whose practices leading to the acceptance of theories are often considered to be the exemplary belief-gaining practices. To illustrate, consider the following to be the list of cognitive values pertinent to accepting theories: accuracy, consistency, predictive and explanatory scope, simplicity and fruitfulness in generating research puzzles (Kuhn 1977). Some have argued that the list ought to include instrumental efficacy (Putnam 1981; or "prediction and control" – behaviorist psychology), high degree of falsifiability (Popper 1959), capability to explain through a narrative what is sound and unsound in historically preceding theories (MacIntyre 1977), inductive derivability and no use of "hypotheses" (Newton: *Principia*), or certainty of fundamental posits (Aristotle, Descartes). Others argue that it ought not include simplicity or explanatory scope (van Fraassen 1980). These disagreements are open to argument; they *are* disagreements, not simply differences in taste. For example, one might reject instrumental efficacy maintaining that it is a social value not a cognitive value (McMullin 1983). Certainty, despite its obvious appeal for ancient and early modern science, has been unanimously dropped from contemporary lists because it is clear that scientific practices cannot produce theories that manifest it; and it has been argued that the explanatory scope is a value derived from pragmatic rather than epistemic interests (van Fraassen 1980). These illustrate the kinds of argument that can be mounted when there is disagreement (details follow in Chapter 5).

In addition to disagreements about items on the list, there can be reasonable controversy about rankings (Kuhn 1977; McMullin 1983; 1996). For example, is explanatory scope more important than simplicity, or accuracy than fruitfulness? There are also disputes about how adequately a particular cognitive value is manifested in a theory. For example, was Copernican theory sufficiently fruitful in view of its weak consistency with the physical theory of its time, or does Skinner's theory of behavior have sufficient predictive power in view of its non-falsifiability (Lacey 1974)? These disputes are open to rational dialogue and the deployment of interpretive skills, though eventually one's answers rest upon practical judgments. In light of these phenomena, it is clear that agreement on a list of cognitive values does not ensure agreement in judgments about acceptable theories. Rationality does not guarantee agreement. This only reinforces the importance of articulating one's cognitive values.

Embodiment of cognitive values in social institutions

In all the preceding respects, the grammar of cognitive values parallels that of values in general, and the parallel continues: cognitive values can also be *embodied in social institutions* and in *society* as a whole. An institution embodies a set of cognitive values to a high degree when it highlights and provides support for those belief-gaining practices that are partly constituted by the set. Scientific institutions, for example, embody some complex of cognitive values containing items like those on the list (p. 53), no doubt with some variation of the items and their interpretations from institution to institution. The variation is within quite narrow limits. None of these institutions embody such once held cognitive values as certainty, and consistency with biblical revelation or with dialectical materialism, or proposals such as intelligibility to the general public or dialectical links with the traditional local knowledge of cultures in the Third World.[5] Since belief-gaining practices require material and social conditions for their pursuit, and since such conditions typically are made available in institutions, the widespread embodiment of a set of cognitive values will be a significant factor in reducing potential disagreements about them, though not necessarily about the beliefs that are held. It can also be a source of ambiguity about the identification of cognitive values, because institutions, even those that exist for the sake of embodying a set of cognitive values (for example, if truth is their motivating value), also embody other personal and moral values and manifest certain social values. Thus, where embodiment is an important factor in there being agreement about cognitive values, it may remain open to investigation to what extent the agreement is to be attributed to the role of these other values (and to considerations of "significance"), rather than to the outcome of "purely cognitive" dialogue and argument.

CRITERIA OF COGNITIVE VALUE: RULES OR VALUES

I have introduced cognitive values as criteria for holding beliefs in general to be consolidated, and then – in order to evoke a sense of continuity between ordinary attitudes of belief and certain stances taken towards scientific theories – taken them also as criteria for accepting scientific theories. But, the extension to scientific theories needs clarification, elaboration and qualification. It might be thought at the outset, however, that granting cognitive values this role in connection with scientific theories is simply to reject outright the view that science is value free, for doing so has values (cognitive values) in play essentially in making fundamental scientific judgments. But the idea of science as value free is concerned primarily with the content of scientific theories and the characteristics and consequences of soundly accepted theories. Thus it depends upon scientific theories being able to be properly *evalu*ated, to be appraised – for their *value* as scientific theories – according to the proper criteria (Scriven 1974). Some scientific theories are better than others and we accept or reject them on the basis of such appraisals; whether (depending on one's epistemological viewpoint) one expresses the "goodness" or the *nature* of the cognitive value of a theory in the terminology of "confirmation," "probability," "corroboration," "verification," "evidential warrant," "verisimilitude" or whatever. The language of value pervades the cognitive attitudes involved in choosing theories. Indeed, I have said that it is a value that science be value free. No irony or paradox is intended. The value desired to be reflected in sound theory choice is not moral, personal, social or aesthetic value, the sort of value from which it is said that science is free. It is cognitive value. The idea of science as value free requires that cognitive and other forms of value be distinct in the sense that answers to "What are the criteria of a theory manifesting high cognitive value?" not depend on answers to "What constitutes a good human life or a good human society?" – that cognitive value be identifiable separately from significance.

Rules

In the preceding section I pointed out that there is dispute about which items should be included in the list of cognitive values. There is also a more fundamental dispute about the character of the criteria of cognitive value for assessing scientific theories. Most agree that the criteria involve certain relations obtaining between theories and available empirical data (and among theories themselves), so that a theory is acceptable to some degree if the appropriate relations obtain. The dispute is about the character of the relations, especially those between theories and data. It has been widely held that (under idealization) the relevant relationships obtain if the relata are connected in a way that can be established by the application of a finite set of formal *rules*, like the rules that (under idealization) underlie mathematical proof; or, if the relationship's obtaining is a matter of degree, that the degree can be calculated by the application of a set of rules. High cognitive value accrues to those theories that can be represented as (or whose high degree of confirmation can be calculated from) the outcomes of (appropriate) formal rule-governed operations. Then, which scientific theories are soundly accepted rests solely on the data, other available theories and the rules. The criteria of cognitive value, appropriate for assessing theory choices, can be represented, so to speak, as governed by formal rules. The intersubjectivity of the data combines with the formal character of the rules to ensure that there can be no place for any kind of values in soundly made theory choices.

Among logical empiricists, the rule-governed character of the criteria of scientific value was often taken as the mark of the objectivity of scientific evaluations. For them, the rules stood in sharp contrast to the criteria of (for example) social value, which could be deployed only in the light of the *judgment* – "subjective" judgment – of a person. Rules, unlike judgments, lead to determinate, unambiguous, non-arbitrary, shared and obligatory outcomes of scientific evaluation. "Objectivity," in this usage, is linked with precision, lack of ambiguity, being able to be measured or calculated, determinateness, and output independent of human variability; and "judgment" with such usages as "it's only a value judgment," and "it's just a matter of judgment, yours against mine," where judgments are held to be subject to endless contestation. "Judgment" can also be used in a more robust sense, as when we refer to the judgment that a theory is the best available one. In this latter sense, according to logical empiricism, sound judgments are the outcomes of the interplay of data and rules, often of calculations of the degrees of confirmation (or inductive support) of theories; so that properly made "scientific" judgments differ in character from "mere" value judgments. In science, it seemed clear to them, high confirmation was the prioritized objective; and the degree of confirmation of a theory was settled by the deployment of rules. Things, however, were not actually this way: not all, for example, accepted that high confirmation was the prioritized objective (Popper 1959); and appeal to rules in fact settled little – in part because there was little agreement about what the rules are – or even about whether they are inductive, deductive, hypothetico-deductive, formalizable within the calculus of probability – and in part because confirmation theory remained underdeveloped. This did not seem to matter: objectivity was linked with rules; subjectivity with values.[6] The contrast, objective/subjective (rational/non-rational) further reinforced that values should be kept out of deliberations about theory choice and theory acceptance. The idea of impartiality has been strongly influenced by the appeal of rule-governed accounts of cognitive value.

Cognitive values

The idea of impartiality, however, need not include that the relevant relationship between theory and data be reflected in rules, but instead can be specified in terms of the appropriate cognitive values being adequately manifested in theories (Chapter 1; see the theses of impartiality stated in Chapters 4 and 10).[7] I neither have a decisive argument that it is impossible to produce rule-governed accounts of the grounds for accepting scientific theories, nor claim that rules play no role at all in concert with the cognitive values. The appeal of rules is obvious enough (and reinforced by cognitive science approaches to cognitive processes), so that on-going efforts, including those that deploy Bayes' Theorem, to overcome the current difficulties facing rule-governed accounts have merit (Salmon, 1983). Nevertheless, the efforts to produce rule-governed accounts seem to me to have remained mired in intractable controversy and stasis; the alternative approach fits well with certain phenomena of the history of science: that disagreement is commonplace in the scientific community; the underdetermination of theory by data, versions of the Duhem-Quine thesis; it fits better with certain philosophical accounts of the nature of scientific theories, for example, the semantic model account; it is more readily applicable in contexts where the relevant research is not exclusively or fundamentally quantitative; and it does not share the philosophical background which denies that values can be the objects of critical, rational discussion; and so it does not link the rejection of rule-governed accounts to the denial of the possibility of objectivity.

Instead of taking the criteria of the cognitive value of scientific theories to be rule-governed, then, I consider them to be cognitive values with all the features laid out in the preceding section.[8] Thus, cognitive values may be manifested in theories, more or less, where the adequacy of their manifestations and their relative rankings and interpretations are matters of reasonable controversy, open to critical dialogue within and with the relevant communities of inquiry and in principle open to "objective" resolution through such dialogue (Hempel 1983a, b; Bernstein 1983; McMullin 1993; Sankey 1997).

COGNITIVE VALUES: A LIST OF COMMONLY CITED ITEMS

I will now review fairly extensively candidates that have been proposed for inclusion in the list of cognitive values. Only on pp. 89–95 (Chapter 5) will I address how one identifies which of these items is and is not a cognitive value, how one settles disputes that may arise about this, and how one distinguishes cognitive values from other kinds of values.

There may not be a single, definitive list of cognitive values. Science is variegated; it admits of a variety of approaches with a variety of aims and forms, and with its historical unfolding new cognitive values may be identified and items, once held to be cognitive values, become reconsidered. The list contains no surprises.[9] Defending its adequacy and completeness, providing a ranking of its items or a general account of how their respective "weights" might be balanced, working out how (and how well) the roles of the various items may be represented in Bayesian models, and even exploring the general consistency of the items – all important matters in need of fuller exploration – lie beyond the scope of this work. I introduce the list here for the sake of concreteness of discussion only; my central theme – the critical evaluation of "science is value free" – does not require a definitive list of cognitive values.

Empirical adequacy. This refers to "the quality of fit between theory and empirical data" (McMullin 1996). What sort of fit? Let T be a theory under consideration for acceptance, and E be the class of relevant empirical data that is available. Minimally, empirical adequacy involves:

1   T is consistent with items of E.
2   T has implications that coincide, to a sufficient degree of *accuracy* with items of E and with statements derived by induction or statistical analysis from items of E.
3   T, via prediction, contributes (or has contributed) to the expansion of E.
4   E includes any unrefuted data fitted by antecedent theories of the relevant domain.[10]

Some clarifying comments: "implications" in (2) does not reduce to "entailments," for typically a theory makes contact with the data only in conjunction with an array of auxiliary assumptions. That is why I have not included *falsifiability* in the account of empirical adequacy or as an additional cognitive value. The formulation of (2) also highlights that often carefully restricted empirical generalizations or data that have been statistically analyzed, rather than individual items or sets of data, are what is brought into contact with theory. Whenever an experiment is replicated, such generalizations, derived inductively, are available where "induction" means either induction by enumeration, by elimination or by curve-fitting.

E refers to the set of available data, items of which must satisfy the requirement of intersubjectivity. The items of E (causally) derive from observation, often of phenomena encountered or created in the course of experimental and measurement practices. A particular theory is expected to "fit" (2–4) only certain classes of data, those derived from specified domains of phenomena. This, though quite obvious, is important to emphasize. We need the "right" data to deal with our theoretical questions; we do not just collect data and then ask which theory fits them. Typically, a theory is held *of a certain domain of phenomena*, or of objects under certain boundary conditions, or of objects insofar as they may be characterized in a certain way; it is not just held in an unqualified way. Empirical adequacy, except perhaps with respect to (1), thus, should be relativized to classes of data (Chapters 7, 9 and 10). The causal origin of data in observation does not imply that putting an item in E is without presuppositions (Chapter 7), that the language used for its description is theoretically neutral or necessarily constant across theoretical developments, or that it may not be revised (or even rejected) in the course of attempts to make T fit with E.

When there is a misfit between T and E, in principle, adjustments may come from either end. Judgments about the appropriate adjustments are made in the light of the full play of all of the cognitive values. What goes into E is neither arbitrary nor (at times) uncontroversial, partly for the reason just stated, and partly because there can be disputes about who is to be included among the subjects of the "intersubjectivity" (Chapter 8). Particular approaches to science (Chapter 5) may add further requirements upon the items of E, or upon the language in which they should be described, for example, by granting special salience to experimental and quantitative data and materialist categories (Chapter 4); to *precision* (*exactness*); or to data that reflect the richness, complexity and variation of ordinary experience (Chapter 9).

Explanatory and unifying power. This is open to two interpretations (Chapter 5): (1) *wide-ranging explanatory power*: explain and unify phenomena (as described in the items of E) in a wide range and variety of domains, perhaps in virtue of the theory offering *reductionist* explanations or more generally because of its explanatory *depth* (Bhaskar 1986); *consilience*: unifying a diverse range of phenomena and of other theories (McMullin 1996, interpreting Whewell); (2) *full explanatory power*: account for all the aspects and the dimensions of phenomena, all their causes and effects, responsive to particularity, concreteness and uniqueness. Clearly this item needs to be elaborated with a full-blown theory of explanation and explanatory ideals.

Power to encapsulate possibilities. Open to a domain of phenomena, including to identify novel, low probability and unrealized possibilities. It incorporates the *predictive power* implicit in the statement of empirical adequacy, and generally implies being able to define the bounds of a theory's predictive power and sometimes having the power to predict *novel* possibilities.[11] Successful prediction reflects the power of a theory to anticipate possibilities within certain bounds or in certain settings. As such it cannot be considered a measure in general of a theory's power to encapsulate possibilities, for this power can be expected to extend beyond these bounds. This item, in combination with the previous one, also incorporates *informativeness* (van Fraassen 1980) or *amount of content* (Feyerabend 1975).

Internal consistency.[12]

Consonance (McMullin 1994), *connectivity* (Ellis 1990), *holism* (Nelson 1995): having appropriate relations with other well established theories, for example, consistency, integration; *inter-theory support*: explaining or being explained by other theories (Newton-Smith 1981); enabling their bounds of application to be specified; and intermeshing through mutual deployment of and dependence on one another's results.

Source of interpretative power: enables the development of an interpretive narrative of the successes and failures of its precursor theories (MacIntyre 1977; Tiles 1985; Taylor 1995) and a specification of the bounds within which they were soundly accepted – sometimes by incorporating (either by explaining or representing as special cases) any of their unrefuted sub-theories; or by *observational nesting* (Newton-Smith 1981), incorporating the observational successes of precursor theories.

Puzzle solving power (Kuhn 1970): the capability to solve both empirical and theoretical puzzles (problems); open to extensions that enable problems to be solved (McMullin 1996).

Simplicity. Notoriously this means different things to different people. In one sense it connotes *harmony*, *elegance* and other aesthetic qualities. I do not take these to represent cognitive values. In another sense it suggests *parsimony*; *economy* (of formulation, of technical devices); efficiency in use for explanatory, predictive and other "scientific" purposes; deployment of the "simplest" available mathematical equations; conceptual clarity, "clearness and distinctness" (Descartes), intelligibility; *idealization* which provides a benchmark, departures from which can be conveniently explained; having appropriate *analogies* with other theories (Campbell 1957)[13] and *formalizability*. In yet another sense it involves *coherence* (McMullin 1994), *appropriate-*

ness, smoothness (Newton-Smith 1981), *epistemic conservatism* (Ellis 1990) or the *rejection of ad hoc features.*[14]

Most of the commonly cited cognitive values appear on my list, but a few do not, for example, *fertility* (fruitfulness, fecundity). It is worthy of more attention, though I am inclined to think that either it is reducible to some combination of items on the list, or it pertains more to such stances as the provisional entertainment of theories rather than to their acceptance (and to the adoption of "strategies": Chapters 5 and 7). My list is intended to provide criteria for the acceptance of theories. Also missing is *inductively derived*; or something like *using only an observational vocabulary* or *not involving hypotheses about unobservables*, as in some interpretations of Newton's "*Hypotheses non fingo*" or in Skinner's (1972) rejection of "theories" (Lacey 1974). This is not to deny the importance of induction in science. As pointed out on pp.58–9, theories often come into contact with empirical data by way of the intermediary of inductively derived generalizations. It is a cognitive value of empirical generalizations to be derived in accordance with sound canons of inductive inference. It is not on the list of cognitive values for soundly accepted theories because, generally, being inductively derived is incompatible with the significant manifestation of such cognitive values as explanatory and unifying power and being a source of interpretative power. Induction also plays a role at a "meta" level. Accepting a theory assumes that it will continue to manifest the cognitive values to a high degree, a meta-inductive inference.

Terms like "being highly confirmed," "being highly corroborated," "having a high degree of verisimilitude," or "true" do not refer to cognitive values. They are cited in attempts to express the *nature* of the cognitive value of a theory; they are not *criteria* whose manifestation in a theory is pertinent to judging its acceptability. The cognitive values are criteria for appraising – depending on one's epistemological orientation – (for example) a theory's confirmation, corroboration, verisimilitude, or truth. Among candidates that have been proposed as cognitive values, that historically have appeared on other lists or that have actually functioned as criteria of theory choice, I definitely exclude the following: *certainty, provable* (as in "mathematical proof"), *consistency with materialist metaphysics, consistency with biblical interpretations, popularity, fitting with "common sense", consensus about what counts as "legitimate" discourse,* and *Baconian utility.*

Baconian utility (*generative of "prediction and control"*; Tiles 1986, 1987) may be considered an ambiguous case, for usually it implies both successful application *and* the service of certain interests. Being of service to certain interests may be endorsed as a social value, but it does not provide reason to accept a theory, as distinct from judging it significant. But that it has been *successfully* applied in (technological) practice is relevant. For successful application of T is *prima facie* evidence that T possesses both explanatory and predictive power in the setting of the application. After successful application of T, its explanatory power has been demonstrated to be more wide-ranging, ranging over applied as well as experimental (and perhaps some natural) spaces. So, successful application (and thus Baconian utility) does not constitute a further cognitive value. Rather, it contributes to the fuller manifestation of some of the cognitive values, for example, the power of T to encapsulate the possibilities open to the objects that have been successfully brought under practical control, but not of others of them. Similar remarks apply to *control*, as it is used in the phrase "prediction and control of behavior," commonly used in the psychological literature. Control of behavior is not an indicator that the theory, which informs the practices of control, encapsulates widely the possibilities open to human behavior (Lacey 1979).

Standards for estimating the degree of manifestation of the cognitive values

Just as personal values may be more or less manifested in a person's behavior, so cognitive values may be manifested more or less in theories proposed for acceptance of various domains of phenomena. Accepting a theory involves judgments that the cognitive values are sufficiently well manifested. How do we estimate ("measure") the degree of manifestation of the cognitive values in a theory? According to what "*standards*" is it measured? If sufficiently demanding standards are not in play, theories will not be submitted to rigorous empirical scrutiny and testing. I suggest that at least the following standards should be in play.[15]

Standards pertaining to empirical adequacy

- *Representativeness*. The items of E are representative of the possible data that could be obtained by observing (often after constructing) phenomena of the domain(s) of which one is considering accepting T. The items do not reflect "bias"; they include data pertaining to all "relevant" characteristics of the phenomena.
- *Pertaining to characteristic phenomena*. The available data include reports of characteristic phenomena of the domain(s). It is important to consider differences and relations among phenomena occurring in various kinds of spaces: experimental, applied (technological), natural, and those of the realm of daily life and experience (Lacey 1984). Phenomena occurring in experimental spaces, for example, may not appro-

priately resemble those occurring in natural spaces (so that, for example, a theory may encapsulate what we can do when we act upon things rather than represent the way the things normally behave). So, attention to this matter is crucial for identifying the domains of which a theory can be soundly accepted.

Applying this standard can be both controversial and value laden. In a famous controversy, Skinner and Chomsky maintained opposed conclusions about the range of domains of which radical behaviorist principles were soundly accepted. At the core of the dispute is Skinner's view that characteristic human phenomena are the most common and generally observed as opposed to Chomsky's that they are unique and distinctive (Lacey 1980).[16] Clearly value judgments are involved. More generally, there are strong dialectical links between considering what are the characteristic phenomena of a domain and judging how it is appropriate to interact with them. This standard is particularly important when considering accepting a theory of a domain of phenomena that are of significance in the realm of daily life and experience (Chapters 7, 9 and 10).

- *Relevant to critical confrontation with competing theories.* E includes data of relevance to the potential rejection of T; data pertinent to putting T into critical confrontation with competitors and to defining clearly the bounds within which T is soundly accepted. This standard (in conjunction with *comparative testability*) involves a generalization of the considerations lying behind Popper's (1959) esteemed proposed cognitive value "high falsifiability."
- *Reliability.* The items of E have been reliably obtained (and they meet the most rigorous standards of numerical *precision* permitted by available measuring instruments), and empirical generalizations derived from them reflect sound inductive and statistical canons. Here a variety of considerations can come into play, including empirical ones.[17] While E does not constitute bedrock, it is the point of reference for accepting theories. Its items derive from experience, but subject to the requirement of intersubjectivity, usually met by replicability. Often, however, the experience is of highly complex, subtle, short-lasting phenomena created by humans with the aid of expensive (and perhaps not well understood) advanced technology, available only to those with specially cultivated observational skills, and difficult to replicate (see Hacking 1983 on the "creation" of phenomena in the laboratory). From whose experience do we obtain the items of E? Which experiential reports should we take as reliable? Often, we are not placed to

replicate them ourselves, and because of expense it may not be possible for any replication to take place. Part of the answer to this involves answering: Whose reports are reliable? This is a question about evaluating testimony and authority (Coady 1992), and the sincerity, trustworthiness, observational acuity, rationality and other virtues of the observer. These issues can become more complicated in the human sciences. Can we expect widespread intersubjective agreement on experiential items regardless of the values of investigators? Or, does observing certain human phenomena and recognizing them for what they are, require a certain moral character or the adoption of certain values? In the natural sciences the notion that one is trained to exhibit the virtues of the "scientific ethos" is widely seen as plausible (Chapter 1). In the human sciences does one have to express certain moral virtues as well? (Taylor 1971; Lacey 1991a, 1997c).

Empirical adequacy is more highly manifested in a theory the more the data it fits meet these conditions, and the more the theory fits the available data that meet the conditions; that is, the fewer anomalies it exhibits.

Generally applicable standards

- *Comparative testability.* T's degree of manifestation of the cognitive values has been "measured" against, and shown to be greater than, their manifestation in a "sufficient" array of (though much less than "all possible") rivals.

Have the "right" array of theories actually been generated to make for a severe test? T comes to be accepted following the unfolding of a research process. In earlier stages of the process, T (or its anticipatory precursors) has been provisionally entertained and adhered to, and these are stances that may be taken towards T in ways that may unobjectionably be partly supported by value commitments. The present standard needs to be met so that any values driving the research process become filtered out, so that the judgment that T is soundly accepted does not tacitly, and without awareness, depend on the possibility that certain rivals were not investigated because conditions for their development were not made available because of their potential links with other values. So, it suggests raising such questions as: have theories that could in principle be applied by those who hold rival value complexes been developed among the array? Do the prevailing social conditions permit the relevant data to be generated (obtained)? It draws explicit attention to the social context of the research process, to the availability of the material and social conditions needed for a research project to develop, and to the values of the participants in the research (Chapters 4 and 10). The matter gains greater importance in those cases where the *neutrality* of T is questionable, and with respect to appraising how well manifested is the cognitive value, "power to encapsulate possibilities" (Chapter 8).[18]

Theories are accepted of particular domains of phenomena. The domains, with respect to which rival theories compete, are also relevant to comparative testability, leading me to propose the following two further standards (both of which figure in Chapter 7) to complement the previous one:

- *Comparative comprehensiveness.* If two theories, T and T' conflict, and if T manifests the cognitive values well of a domain D and T' manifests them well of D', where D' is contained in D, then *ceteris paribus* T manifests the cognitive values to a higher degree than T'; even more so if, from its perspective, the success of T' with respect to D' can be explained.

- *Comparative local strength.* T manifests the cognitive values highly of D, where D has sub-domains $[D_1, \ldots, D_n]$ only if for each sub-domain D_i it manifests them more highly than any competing theory of D_i.

Finally, in addition to comparative testability and its complements, I also include the following:

- *Comparison with the most well entrenched theories.* The manifestation of the cognitive values in T compares favorably with their manifestation in the most soundly entrenched theories, those considered to belong to the stock of knowledge.

- *Response to criticisms.* Criticisms that T does not adequately manifest the cognitive values have been adequately responded to, especially criticisms that make explicit what would count as more adequate manifestation. The degree of manifestation of the cognitive values increases under conditions of strong criticism.

Clearly these (and any other) standards will have to be deployed within a mode of interpretive understanding, related to the context and practices of generating and collecting empirical data and of developing and testing theories. Their being deployed adequately may rest upon certain social conditions obtaining, upon the personal values of participants in scientific practices and upon the diversity of values held among these participants. Moreover, the push to more rigorous interpretation of the standards may come from those who adopt particular value complexes ("Rudner's argument": Chapter 4) for they – seeing the acceptance and practical adoption of certain theories as contrary to their interests – may sometimes (though other times the opposite) more rigorously evaluate the credentials of supposed authorities, and seek out the conditions that would enable a wider array of theories to be developed and tested.

The key idea of impartiality maintains that soundly accepted theories are those that manifest the cognitive values to a high degree as estimated in accord with the most rigorous available standards. We turn now to the discussion of this idea.

4   Science as value free: provisional theses

The sources of the idea that science is value free have been discussed (Chapter 1). Drawing from them I will now motivate and state provisional theses of impartiality, neutrality and autonomy, which will later be criticized and revised (Chapter 10).

IMPARTIALITY

Impartiality is a view about properly accepted theories. A theory (T) is accepted of a domain (D) of phenomena, things and their possibilities.[1] To accept T (of D) is to judge that it is so well supported that it needs no further investigation, that it should be included in the stock of consolidated beliefs. The core of *impartiality* is that T is "soundly" or "properly" accepted only if such judgments are based solely on assessments of the cognitive value of the theory, regardless of any considerations of significance; so that to accept T (of D) is (ideally) to make the judgment that believing T (of D) has high cognitive value. The cognitive values are the criteria of cognitive value. Most of them involve relations between theories and available empirical data and relations among theories. The possibility of *impartiality* depends upon cognitive values being distinct and distinguishable from [other] values.

The seemingly innocuous phrase "available empirical data" hides various complexities and potential controversies. For convenience I will use the term "empirical datum" to denote both an observational report and what is reported in an observational report (an "observed fact"). Not all empirical data of phenomena of D are relevant for the assessment of T. For a theory of projectile motion, for example, the height the arrow reaches in its flight is relevant but not its color. But we cannot "measure" the degree of manifestation of the cognitive values in T until the class of empirical data to which T is expected to be related in the specified way has been *selected*. How is the relevant class of empirical data selected?

Materialist strategies

The Baconian idea (Chapter 1) introduces intersubjectivity as a condition, and consequently grants high salience to experimental data. From the Galilean idea (Chapter 1) follows the further restriction that the descriptive language in which the data are expressed contain only what I will call "*materialist*" terms: generally quantitative and mathematical, applicable in virtue of measurement, instrumental intervention and experimental operations – the kinds of terms that apply to phenomena considered as generated from underlying structure, processes and laws rather than considered as an integral part of daily life and social practice. Limiting the lexicon of the relevant data to contain only materialist terms is a much stronger condition than intersubjectivity. The phenomena, described by the data, can be characterized in indefinitely many ways. Many observed phenomena described with non-materialist terms (like many described only with materialist ones) can be grasped intersubjectively in observation, for example, reports of actions described with intentional terms. The Galilean idea leads us to select data that describe phenomena abstracting from their place in any human practice or their relation to human experience; data that have been stripped of all links with value.

The data are selected in this way because they are the kind of data that can bear as evidence upon theories that have been provisionally entertained. Corresponding to this condition on the selection of data, then, there is a *constraint* on the kind of theories that may be provisionally entertained: that theories only deploy categories capable of representing the appropriate kind of underlying structure, process and law. Thus, for example, theories that deploy teleological, intentional or sensory terms are not to be entertained. Prior (logically, not necessarily temporally) to any consideration of the acceptability of a theory, following the Baconian/Galilean ideas, what I call the *materialist strategies* are in play. Developing theories under the materialist strategies[2] makes it possible to bring theories into contact with data in such a way that the degree to which they manifest the cognitive values can be "measured."

Later (Chapter 5), I will propose that the materialist strategies constitute just one among several kinds of strategies that can be adopted while continuing to emphasize the intersubjectivity of empirical data. Then (in Chapter 6) I will discuss more fully the materialist strategies and their grounds,[3] and reject the view that the referents of materialist terms have a special ontological status. Influenced by Baconian and Galilean ideas, however, most of modern science proceeds under materialist strategies. Note the plural: "strategies." Modern science admits of considerable variety. Some of it emphasizes systematic description rather than explanation; some structure rather than law; some phenomenological or statistical regularities rather than underlying law; some finding law through inserting an object into a larger, encompassing structure. Not all science is conducted under "physicalist" strategies: constraining theories to be open to reduction to physics, or deploying categories derived from physical or chemical theory – though reductionist tendencies are widespread – and where they are not present it tends to be a requirement that theories be left open in principle to eventually being subsumed under laws in conjunction with posits about underlying (possibly developmental) processes and (possibly emergent) structures.[4] Sometimes materialist strategies may include further constraints, (for example) restricting laws to being determinist, processes to being mechanical, and (in behaviorist psychology) the independent variables of laws to being instantiated by environmental events. Materialist strategies exhibit multiplicity and variability.

The materialist strategies indicate the kinds of data that are needed in order to estimate the degree of manifestation of the cognitive values in a theory (of a domain), and to which any theory which is a candidate for acceptance of the domain is to be held accountable. Following the strategies one seeks to obtain the data, constructs the experimental phenomena, the observation of which can occasion gaining relevant data, makes the relevant observations and records their outcomes. Both logically and temporally, the class of available data is open-ended, and it never exhausts the class of possible relevant data. What data are available varies with time; and individually and collectively, in principle, they are subject to revision, even (rarely in practice) fundamental reconstitution. Which ones are available at a given time, especially concerning experimental phenomena, is a function of our interventions and foci of interest, and often of available technology or resources. Because of these contingencies, the data available at a given time (despite being of the required kind) may not enable a judgment to be made about the cognitive value of a theory and its competitors; then judgment should be suspended pending further investigation.

A thesis of impartiality

In the light of the preceding considerations, I offer the following provisionally as a thesis of impartiality:[5]

I′   1   The cognitive values are distinct and distinguishable from other kinds of values.

2   A scientific value (T) is accepted of a domain of phenomena (D) if, and only if, T (of D) in relation to the available empirical data (E) manifests the cognitive values highly according to the most rigorous available standards; and to a greater degree than does any rival theory – where T meets the constraints of, and the items of E have been selected in accordance with the materialist strategies.

3   T is rejected of D if, and only if, another theory (T′) is accepted of D, and T and T′ are inconsistent.

Hence:

4   Values, and assessments of a theory's significance, are not among the grounds for accepting and rejecting theories.

5   Only the materialist strategies generally constrain theories that may be provisionally entertained and put requirements upon the kinds of data selected.

I′(2) is the key component. Is it a fact that theories in modern science are accepted in accordance with it? No doubt it often is, but there are plenty of exceptions (Chapter 9), cases in which agreement among members of the scientific community to accept a theory is explained only by their sharing particular values. It is easy to understand how there can be such exceptions, for it is consistent with (5) being an articulated value of scientific practice that particular members of the scientific community – having an interest (consciously or not) in consolidating theories that are significant for their adopted value complex – may actually employ further constraints on the kinds of theories they wish to come to accept (and reciprocally on the kinds of data they select). According to I′ adopting these further constraints is not a ground for rejecting (denying the cognitive value of) theories that do not fit them. It may be a good reason, however, to deem these theories lacking in significance. Accepting a theory does not imply its significance; and deeming a kind of theory to be lacking in significance does not imply that theories of that kind may be rejected. It may sometimes be a good reason not to investigate them; but at other times not to investigate them leads to accepting theories without their having been tested against a sufficient range of competitors. Vigilance is necessary to avoid slipping from insignificance to rejection, and then from rejection of its rivals to acceptance of a theory.

The existence and intelligibility of exceptions to (2) and (3) do not seem commonly to lead to the denial of these items among members of the scientific community. They seem to function as an ideal; so that "accept" is taken to connote "soundly accept" or "properly accept." The exceptions tend to be seen as lapses, not as characteristic of occasions of accepting theories; and, when they are clearly identified, acceptance is usually withdrawn. I'(2)/(3) expresses a value of scientific practice, a value whose manifestation is diminished by departures from it in fact – and (1) is a presupposition of its adoption. Those who adopt this value consider it a regulative ideal of scientific practice that all theoretical conflicts can be resolved in accordance with (2)/(3) consequent upon suitable growth and transformation of the available empirical data and the rival theories, and they affirm that throughout the modern scientific tradition many conflicts have been so resolved – leaving us with a solid and expanding stock of soundly accepted theories.

Rudner's argument

When we accept T (of D) we never gain certainty. Nevertheless accepting T (of D) in accordance with I'(2) carries with it (pragmatic) certitude, well vindicated by the great successes of the practical application of the best accepted theories of modern science. Pragmatic certitude is not the same thing as epistemic certainty; but when we have certitude of T (of D), it seems clear that no cognitive barrier remains to its application. Even so, in all sorts of unanticipated ways the body of available data may change or a hitherto unthought of theory might pose a challenge so that it always remains possible that T (of D) may come to be rejected. Were it to become rejected, we might expect that there would be consequences of having practically applied it that would have negative moral value.

In a famous article, Rudner (1953) argued that when we accept T (of D) we are committed to the judgment (which I paraphrase into my terminology and call "Rudner's condition"):

> (R) T manifests the cognitive values to a sufficiently high degree (of D) so that the legitimacy of its being applied in practical activity is not to be challenged on the ground that, if T (of D) were to turn out to be false, consequences of negative moral significance (consequences undesired for one's value complex) might follow from so applying it.

We might put it: we need sufficient certitude to compensate for our necessary lack of certainty; but gaining that certitude is implicated in the value judgment that the undesirable consequences of applying T in practical activity T, should T actually be false, do not warrant (pending further investigation) withholding on the application of T. We need sufficient certitude to compensate for morally significant risks that would be taken by applying a false theory in practice. Rudner goes on: "The scientist, *qua* scientist, makes value judgments." If he is right, denying (R) concerning T (of D) seems to imply denying I'(2) concerning it. Does not (R), therefore, conflict with (4) – so that I' does not even express a coherent ideal that could be embraced as a value?[6]

Can we affirm I'(2), but deny (R)? For what type of reason? It would have to be that the cognitive values were not sufficiently manifested in T (of D). But, by hypothesis, T manifests the cognitive values to a very high degree according to the highest recognized standards of evaluation – so much so that the scientific community concurs that further investigation *ceteris paribus* would serve no cognitive interests. So, the denial of (R) would involve questioning the stringency of the recognized standards of evaluation. For example: Was T adequately tested against data that are representative of characteristic phenomena of the domains in which practical applications of T would have consequences, in particular the ones that would be undesired should T be false? Was T tested adequately against rival theories which, should they be true, would involve fewer such undesirable consequences if adopted? Is the community of scientists adequately constituted to warrant deference to consensus achieved in it? In principle, these questions can always be raised because the domains of testing (especially experimental) and application do not coincide, because theory is underdetermined by the data, and because the criteria of what constitutes scientific competence may implicitly be tied to particular values. I am interested in the questions here, however, only insofar as they lend themselves to answers in practice, for example, when the alleged characteristic phenomena are specified, a rival theory outlined, or evidence presented that the composition of the scientific community makes it prone to "bias"; for instance when they point to inquiry relevant to answering the questions.

Questioning the standards in this way implies that there are not adequate grounds to accept T (at least of the domains relevant when considering its practical application). Thus, the same ground that serves to deny (R) also serves to deny (2). Then, *ceteris paribus* affirming (2) implies affirming (R). Accepting T (of D) implies the judgment that all the testing relevant for affirming (R) has been conducted; so that it involves making both of the judgments (2) and (R). The key link between them is provided by the standards for "measuring" the degree of manifestation of the cognitive values in a theory. A value judgment may be appropriately, and essentially, involved in assessing these standards. Thus, holding a particular value complex may lead one to push for higher standards; and differences in the value complexes held – and thus possibly different judgments regarding (R) – may lead to different standards being deployed, and thus different judgments regarding (2) actually being made. But the dispute in these situations is not about I'; it is about the standards. In any case, the different judgments are "to accept" and "not to accept." If one makes the former, one judges that T (of D) manifests the cognitive values highly according to the most demanding available standards; it is irrelevant to making that judgment that values may influence our sense of what are sufficiently demanding standards. If one makes the latter, it remains that no theory has been accepted or rejected with values among the grounds; the values push one to adopt more rigorous standards and thus to engage in further investigation, not to reject the theory. There is, therefore, no conflict between (4) and (R). Rudner's argument does not threaten I'.

While Rudner's argument does not threaten *impartiality*, and so does not imply that values and science "interpenetrate," it does point to important aspects of the "touch" (or shall I say "constant rubbing") of science and values – referring back to Poincaré's metaphor (Chapter 1). Holding certain values may push in the direction of adopting more demanding standards for estimating the degree of manifestation of the cognitive values; so that *impartiality* may be furthered more constantly within some value complexes than others. Where a value complex underlies making the value judgment of denying (R), it may open up fields of empirical inquiry effectively closed by those who affirm (R) – perhaps without their being aware of doing so or perhaps because they are driven by economic interests and they cannot market the products of a discovery while (R) is denied; but it does not put constraints on the content of acceptable (as distinct from significant) theories.

I emphasized "practical" questioning of the standards, questioning that points explicitly to further (novel, not simply replicative) inquiries, based on the proposed more demanding standards, which (depending on the outcome of the inquiries) could vindicate the point of the questioning or put an end to it.[7] Pending the outcome of the inquiries, it is appropriate to deny (R) and therefore (2); even though – by the consensus of the scientific community – according to the hitherto recognized highest standards (2) would be affirmed. The very fact that the standards are being called into question breaks that consensus (at least temporarily), and the very fact that the proposed standards are more demanding is *prima facie* reason for adopting them. Here, several difficulties can arise. In the first place, practical, material or social conditions may not be available to carry out the inquiries, or may actively pose barriers to them. The higher standards may be recognized (by some), but not be effectively available in practice. Under these conditions, it will be appropriate to continue to deny (R), and therefore (2).[8] Second, the new standards may be recognized by some, but not by others (perhaps even the majority of the scientific community). They may not be understood; this is not implausible if the new standards derive from reflections on what are the characteristic phenomena of the domain of intended application (especially in the human sciences), and if the participants in the scientific community represent a limited range of value complexes. Or, they may be disputed and the dispute may not be resolved. Concerning some domains of inquiry, consensus regarding the standards may not be reached. In such cases we cannot expect that theories will be accepted in accordance with I'. That would not be a ground to reject *impartiality* as a value, but only to recognize that the conditions for its realization are not always available. In these cases we only have the pulls and pushes of opposed opinions (where what is held to be possible tends to be derived from what is desired, and where acceptance of theories tends to be confused with judgments of their significance) with power as the ultimate "authority," and where the possibility of *neutrality* does not arise.

NEUTRALITY

Impartiality concerns the sound acceptance of scientific theories and its grounds. In this section, for the purpose of stating a working provisional thesis, I will consider *neutrality* to concern the logical implications and various consequences of accepting theories.

Neutrality is important to the self-image of the modern scientific tradition. Think, for example, of the rhetoric of science as a universal community, a universal culture, and a unifying humanistic force cutting across barriers of culture, race, religion and gender, generating theoretical products that are of benefit to all and threatening to none (Bronowski 1961). Regardless of culture, race, religion, gender – moral and social values in general – science (its accepted theories and its practices of inquiry) is always an object of value, so that one does not have to resolve value disputes in order to value science. Science, it is said, has (or aspires to manifest) features in virtue of which, for every value complex, it is an object of value; it belongs to the common patrimony of humankind; it is a public good.[9] I will state a thesis that attempts to express what those (aspired to) features of science are, building upon the three components to the idea of neutrality (introduced in Chapter 1). The rhetoric also suggests that not only is science an object of value for all value complexes, but also that it is an object of value partly *because* it is *neutral*, not just because piecemeal it serves some particular interests of each value complex. This suggests that, for any value complex, not only is science an object of value but also the value of *neutrality*, articulated in a defensible coherent thesis of neutrality, is (should be) endorsed: science is *neutral* and *neutrality* is a universal value. We will see on pp. 79–82 that it is difficult to hold both.

The three components to the idea of neutrality (Chapter 1), adapted to the context provided by the proposed thesis of *impartiality*, may be put as follows:

1   "Consistent with all value judgments": Any value judgment is consistent with a scientific theory accepted in accordance with I′ – more generally, accepting a theory in accordance with I′ implies (logically or rationally) no commitments about values to be held.
2   "No value consequences": Accepting a theory in accordance with I′ has no consequences for the fundamental values one holds; it neither supports nor undermines the holding of a value complex.[10]
3   "Evenhandedness in application": For any value complex, theories, accepted in accordance with I′, may be applied in principle to significant phenomena and in practical activities in ways that further (and do not undermine) the interests grounded in the complex; or: For any value complex, theories accepted in accordance with I′ are significant to some non-trivial extent – in principle, they can be put at the service of any values, explaining valued phenomena, illuminating the realm of the possible, informing means to ends and the attainability of ends.[11]

"Consistent with all value judgments" is true; it has been "made true" by the adoption of the materialist strategies to represent phenomena in abstraction from all contexts in which values are manifest. The other components confront serious difficulties. "No value consequences" does not follow from "consistent with all value judgments" (see p 77); and to defend "evenhandedness," it seems that considerable emphasis has to be put on the "in principle," for modern scientific knowledge has been practically applied far more readily for the sake of advancing technology to serve economic, military and other projects in the advanced industrial world than it has been in association with any other values (cf. Harding 1998).

The idea of neutrality, with these three components, clashes with another facet of the self-image of the tradition of modern science – science's special service to *progress*. Progress is not neutral; it cannot coexist with the traditional values of numerous cultures. This latter facet of modern science's self-image often celebrates the falsity of the "no value consequences" component, as in the following classic passage:

Science as an institutionalized art of inquiry has yielded varied fruit. Its currently best-publicized products are undoubtedly the techno- logical skills that have been transforming traditional forms of human economy at an accelerating rate. It is also responsible for many other things not at the focus of present public attention, though some of them have been, and continue to be, frequently prized as the most precious harvest of the scientific enterprise. Foremost among these are: the achievement of generalized theoretical knowledge concern- ing fundamental determining conditions for the occurrence of vari- ous types of events and processes; the emancipation of men's minds from ancient superstitions in which barbarous practices and oppressive fears are often rooted; the undermining of the intellectual foundations for moral and religious dogmas, with a resultant weakening in the protective cover that the hard crust of unreasoned custom provides for the continuation of social injustices; and, more generally, the gradual development among increasing numbers of a questioning intellectual temper toward traditional beliefs, a development frequently accom- panied by the adoption in domains previously closed to systematic critical thought of logical methods for assessing, on the basis of reli- able data of observation, the merits of alternative assumptions con- cerning matters of fact or of desirable policy.

(Nagel 1961: vii)

As an unrestricted generalization it is false that accepting a theory has no consequences for the fundamental values one holds. In some historically striking cases, alluded to in the quotation, the consequences of accepting a theory have indeed included undermining certain fundamental values.

Accepting a theory and consequences for value judgments

The Galileo case is perhaps the most celebrated in which a theoretical development had among its consequences the undermining of traditional values. We may summarize the logic underlying this case as follows. In Aristotelian science, the domain of the planets had two descriptive frameworks: one composed of geometric/kinematic categories, the other relating the planets to the cosmos in virtue of which their arrangements and movements were explicable teleologically. The centrality and immobility of the earth figured prominently in both frameworks. The teleological account was directly relevant to a value complex. In medieval ideology the order of the cosmos grounded the predominant value complex, especially that pertaining to the social values which were to be expressed. The grounding goes roughly:

1   Proper social relations, those which manifest the desired values, are God-ordained.
2   God-ordained social relations mirror the order of the universe.
3   The hierarchical order of the Aristotelian cosmos is the best account of the order of the universe.
4   The Aristotelian cosmos reveals the earth as stationary and central.
5   Therefore, the earth is stationary and central.
6   Feudal social arrangements mirror the hierarchical order of the Aristotelian cosmos.
7   Hence, in the absence of alternative candidates, feudal social arrangements are God-ordained.
8   Thus, it is impossible to transform the feudal order into a better order.

A crucial part of the argument for (3) rested on the theoretical context and empirical evidence for (4). With the refutation of (5) (see Chapter 7), (3) is also refuted and the argument for (7) dissolves, since no alternative grounding of the feudal value complex was able to be constructed. In this way a judgment of Galilean science is incompatible with the feudal value complex, being inconsistent with its presuppositions, though not with any of its value judgments. According to *impartiality*, the feudal value complex must adjust. (Confronted with the incompatibility Galileo's persecutors opted to reject *impartiality*.) In the process, the argument for (8) also dissolved, and with that dissolution came the removal of a barrier to pursue the interest of extending human control over nature (which would necessitate new social relations) rather than remain confined by obligations to retain the old social order (which was implicated in a stance of adaptation or attunement with nature, Chapter 6). Thus, scientific developments not only removed a pillar of the old value complex, but also were part of the process that would show that a new value complex could become manifested. The Copernican revolution achieved the former; the acceptance of theories confirmed experimentally achieved the latter by displaying that the possibilities of control over nature far outstripped previous anticipations (Chapters 6 and 7).

I have cited an historical example, as well as an aspect of the self-understanding of the tradition of modern science, to cast doubt on "no value consequences." It might seem that "consistent with all value judgments" implies "no value consequences"; then doubts about the latter would suffice to also cast doubt on the former.

But it does not. I have maintained that a value complex consists of an integrated ensemble of values and value judgments rendered coherent by various presuppositions about human nature and about what is possible and not possible. Since a scientific theory provides encapsulations of the possibilities allowed by a domain of phenomena (Chapter 5), it may imply to be impossible (possible) what a value complex presupposes to be possible (impossible), or it may contradict a posit of a conception of human nature. So theories may be inconsistent with the presuppositions involved in adopting value complexes. Suppose that T, a theory accepted in accordance with I′, is inconsistent with a presupposition (p) of adopting a value complex (c). (For example, p may be (5) in the Galilean example, and c the feudal value complex.) Then, accepting T implies: 1) *cp* [*ceteris paribus*] rejecting p, and thus 2) *cp* ceasing to adopt c which (in turn) will lead 3) *cp* to the rejection of c. In this sense, accepting T undermines or is incompatible with adopting c; though, reflecting the *cp* in (2) and (3), since c may be newly rendered coherent under different presuppositions (p′), T remains formally consistent with the value judgments contained in c.[12]

Revision of "no value consequences"

The growth of scientific knowledge, the consolidation of more and more theories in accordance with *impartiality*, thus, can lead to challenges of value complexes (and thence their social sustainability) simply by undermining key presuppositions of their legitimation. *Impartiality*, therefore, actually challenges a key component of the common idea of *neutrality* rather than being the opposite side of the same coin; though for whatever measure of *neutrality* there is achieved, *impartiality* constitutes a necessary condition. Moreover, if "no value consequences" is false, it would not be surprising if there is little room to apply the theories (accepted in accordance with I′) in practices valued from the perspective of threatened value complexes, at least without the applications becoming the occasion for these value complexes to become more threatened; and so "evenhandedness" would also be false.

Thus, it is not plausible to think of *neutrality* as cutting across all value complexes, but only across the "viable" ones. Rather, we might take *impartiality* − I′(2) and (3) − to define the limits of application of *neutrality*. I will say that a value complex is *viable* if its presuppositions are consistent with the body of theories which have been accepted (of the relevant domains) in accordance with *impartiality*. Some value complexes are not viable, as Nagel celebrates. But there remains an array of viable value complexes, the array of modern pluralism. When we limit consideration only to this array, accepting scientific theories neither supports uniquely nor undermines any value complex. Theories pertain to values only where they illuminate the workings of significant phenomena and deliberations about means to ends and the attainability of ends; and in principle, and to a high degree in practice, do so regardless of what the values are. To its proponents, *neutrality* pertains among the items of this array. I will drop the "no value consequences" component in favor of the "range of viable value complexes" component. The "evenhandedness" component of *neutrality* can now be restated: For any viable value complex, theories, accepted in accordance with I', (in principle) are significant to some non-trivial extent.

A thesis of neutrality

Against this background, retaining "consistent with all value judgments,"[13] transforming "no value consequences" into "range of viable value complexes" and restating "evenhandedness in practical application," I offer provisionally the following statement of a thesis of the neutrality of scientific theory and theory acceptance:

N′ 1   No scientific theory (accepted in accordance with I') has value judgements among its logical implications; and accepting a theory in accordance with I' implies rationally no value commitments.

    2   There exists a range of *viable* value complexes; then, accepting theory neither undermines nor supports the holding of any viable value complex.

    3   A theory accepted in accordance with I', in principle, is significant to some non-trivial extent for any viable value complex.

N′ states a modest thesis, some would object too modest and that, in particular, it does not capture well enough the way in which the applicability of theories is evenhanded. It is not just, the objectors would say, that for any viable value complex an accepted theory (in principle) is significant (but with possible great variations of significance across the complexes), but that (in principle) all viable value complexes are equally able to gain from the application of accepted theories in practical activities, or that accepted theories are available to be applied to serve the good of everyone, regardless of their adopted value complexes. If such a stronger version of "evenhandedness" could be established generally, it would follow that gaining and adopting scientific knowledge would be itself a (social) value within all viable value complexes – a view, as pointed out at the beginning of this section, widely associated with *neutrality*. Let us, then, consider replacing (3) by (3a):

3a   Every viable value complex gains, in principle, a fuller realization from the practical application of theories accepted in accordance with I′, and none are better suited than others to apply them for the sake of furthering their own realization.

Thence:

3b   In principle, the interests of some value complexes are not especially furthered, rather than those of others, by the gaining and application of accepted scientific theories.

The difference between (3) and (3a) is important: (3) requires only that, for each accepted theory, application is possible (in principle) within the context of any viable value complex. Compatible with that, in practice, the extent of applying scientific knowledge (and the capability to apply it more extensively) may vary markedly within the contexts of different value complexes – from application pertaining principally to marginal (or isolatable) activities to it shaping the central productive acts and contexts of daily life and experience. In the latter cases *de facto* the application of scientific knowledge has a more important role in the social implementation and consolidation of the value complex. Often, this will be not merely *de facto*, but also "in principle," for in the former cases the more extensive application of scientific theories may require conditions, and generate consequences, that undermine the value complex. For certain (many?) particular theories, (3a) will be defensible, but some complexes are generally more suited than others for the application of theories, contrary to (3b). Compare the role of the practical applications of theories (accepted in accordance with I′ and thus developed under materialist strategies) in a Latin American peasant culture with those in advanced industrial countries. Both provide instances of viable value complexes, yet clearly the one whose central activities and experiences are informed by scientific knowledge is more suited to apply scientific theories than the one whose central activities are not.

In the extreme case a value complex may be viable (consistent with theories accepted in accordance with *impartiality*), but not *historically and socially sustainable*, for instance the social and material conditions necessary for its component values to be manifested are not available and its central practices, largely uninformed by scientific knowledge, are unable to resist the thrust and expansion of "development" practices which are well informed by scientific knowledge (Chapter 8). Moreover, wherever there have been clashes between modern and (viable) pre-modern or grassroots alternative value complexes, applied science has always served the interests of the modern. (3) does not imply (3a), and (3a) represents neither a fact, nor even an idealization of historical fact. If (3a) represents a value, it is far from being highly manifested in the practices of modern science, and it is not endorsed by those who hold certain viable value complexes (Chapter 8). So I will not include it in the statement of N'.

I am attempting to formulate a clear thesis in which *neutrality* cuts across viable – not socially sustainable – value complexes, across those value complexes that are cognitively admissible not just across those permitted by current arrangements of power. In doing so I wish to remain faithful both to the initial idea of neutrality and to the historical record that scientific advances have had among their (cognitive) consequences the undermining of certain traditional value complexes, and thus to be able to reconcile endorsing both *neutrality* and progress. The upshot has been that the proposed thesis, N', does not capture a robust sense of "evenhandedness" across value complexes. Perhaps I have strained too much to be inclusive by defining "viable" only in terms of consistency with theories accepted in accordance with *impartiality*. Others may want to define it more stringently. *Neutrality*, it might be proposed, is across value complexes that include the value of "rationality," supposedly a universal or a cognitive value, where rationality requires not simply consistency with soundly accepted theories, but also *engaging in practices of central importance to society in a way that is informed by theories accepted in accordance with I'*. Restricting "viable" to those value complexes that include rationality understood in this way, while it enables the expression of a robust kind of evenhandedness of application, does so at the price of maintaining that *neutrality* holds only across value complexes that include the high salience of a particular social value (not a cognitive value), namely choosing one's actions and practices so that they can be informed by knowledge gained under materialist strategies.

Still others might wish to restrict *neutrality* to hold across those value complexes whose presuppositions are consistent with materialist metaphysics, not just consistent with theories that are accepted in accordance with I' and thus accepted under materialist strategies. This proposed restriction, like the one in the previous paragraph, generates a kind of paradox. The paradox arises because the best grounds for accepting materialist metaphysics are linked with adoption of "the modern values control" (Chapter 6), so that to restrict viable value complexes in this way is equivalent to restricting them to those that include the modern values of control. Then, neutrality across values holds only where particular values are held. This tends not to be seen as paradoxical because the adoption of the modern values of control is widely taken to be universal, an essential part of the self-definition of modernity, where not to adopt them is seen as somehow to misconstrue the way the world is and to be maladaptive to finding one's way around in it.

Later (Chapter 10), I will consider an alternative to N′ (N) in which (3) will be replaced by:

> 3′   for any viable value complex, in principle, there are some theories accepted in accordance with I′ that are significant.

(3′) is a weaker thesis than (3). It does not imply that any accepted theory can, in principle, serve interests of all viable value complexes. One may endorse (3′), yet not hold the view – that partly motivated consideration of (3a) – that in general the unqualified gaining of scientific knowledge is a value. We do not get "evenhandedness" in a robust form from (3′): not every theory accepted in accordance with I′ (and thus under the materialist strategies) need be applicable in every (viable) value complex; but for each (viable) value complex there are theories accepted in accordance with I′ that are significant to some degree. Finally (Chapter 10) I will introduce a further version (*N*) which I will endorse. This version of *neutrality* is not restricted to the context of research conducted under the materialist strategies. It attempts to capture a measure of "evenhandedness of application," and expresses *neutrality* as a thesis about the character of scientific inquiry rather than simply about the implications and consequences of accepting theories in accordance with *impartiality*.

AUTONOMY

The idea of autonomy concerns features of the processes and practices of science that are conducive to generating theoretical products that manifest *impartiality* and *neutrality* to a high degree; so that it is subordinate to them. It proposes that the processes of science are driven by purely cognitive considerations, by the goals of collecting more empirical data and developing more and more encompassing theories that become accepted in accordance with I′, and not by "outside interferences" that are irrelevant to the sound acceptance of theories. Among "*outside interferences*" I include: values and (for example) metaphysics, power, personal ambition, popular appeal, government, economic interests, law, religion, the military, ideology, the "will of the majority" and special interests of any kind.

Autonomy is a value – of course not always (or even often) manifested. It is often supported by pointing to historical cases in which outside factors have in fact interfered with the process of science, and where the process, aiming to gain products that accord with *impartiality* and *neutrality*, has been distorted, threatened, decelerated or compromised. The idea of autonomy may also build in the further proposals that outside interferences are more likely to be discerned and resisted if the members of the scientific community have been formed in the practice of the "scientific ethos" (Chapter 1), and if scientific research is conducted by an "autonomous" ("self-directed") community working within "autonomous" institutions.

I offer for consideration the following thesis designed to capture the core of the idea of *autonomy*:

A′ 1   Scientific practices aim primarily to gain more, deeper and more encompassing theories that are accepted in accordance with I′ and whose acceptance accords with N′.

2   They are conducted without "outside interference" by the scientific community which, in order to ensure the furtherance of (1): a) defines its own problems, questions, priorities, and domains of phenomena to be investigated; b) has unique authority with respect to matters of method, theory acceptance, and standards (both cognitive and moral) of scientific conduct; c) determines who, and the qualifications of whom, will be admitted into the scientific community, and what counts as competence and excellence; d) shapes the form and content of scientific education, and the structure and activities of scientific institutions; e) forms its members in the practice of the "scientific ethos"; and f) exercises its responsibility to the public fully by acting in accord with items a)–e).

3   The scientific community conducts its investigations in self-governed institutions which are free from "outside interference", but provided with sufficient resources in order to conduct its investigations efficiently and reliably.

A′ is riddled with tension. Scientific institutions and research practices require financial, material and social conditions which they do not generate by their own activity, and so they must be provided by "outside sources" (generally government, business and universities). There cannot but be "outside influences" on scientific institutions; *autonomy* depends on curtailing the "influence" so that it does not become "interference." How can this be done? There are not unlimited resources, and provisions for scientific research must compete with other legitimate interests. It is properly a matter for public deliberation, not an internal scientific question, to consider how scientific research (and in what areas) ranks in importance with such other social values as alleviating hunger, maintaining economic growth, improving education and sustaining the environment.

Whatever balance is reached, the outcome will involve limits upon the resources available for research, and require adjustment of its direction and emphases. Thus, it appears, "compromises" must be made with (2) as well as (3). Can they be made in such a way that the outside "influences" play along with (1) and not in opposition to it?

Any compromise will involve adjustment to some of the items of (2). An "acceptable compromise" might qualify items (a) and (d) of (2) (and only them) by adding "in collaboration with the appropriate 'outside influence' – government agency, biotechnology company, etc." to the beginning of each of them, and "where the collaboration is not to such an extent that it involves a practical denial or subordination of (1)" to the end. Scientists cannot investigate everything; there is always a choice about which phenomena and domains of phenomena to investigate, and about the priorities of research. To some extent no domain is devoid of scientific interest, although clearly the value represented by (1) is better served by engaging in research in selected domains. But investigating any domain can contribute to expanding the compass of scientific knowledge. There is, then, some arbitrariness (even a considerable amount) while remaining consistent with (1) in choosing what domains to investigate, leaving plenty of room for the choice to be influenced by personal and institutional values and outside factors. So long as the influence of these factors is limited to that of domain of investigation I' will not be threatened, for choices made in accordance with I' are about theories of specified domains. So long as some space is left open for "fundamental research," that directly concerned with the gaining of deeper and more encompassing theories, that outside factors play some role in these choices may *de facto* not subordinate (1) to their goals.

Such compromises, it may plausibly be argued, are necessary for the progress of science. Without them the conditions needed for the conduct of scientific practices will not be provided, so that without them there can be virtually no progress of scientific knowledge. With them, however, theories can come to be accepted of certain domains of phenomena, even if (often?) the domains are chosen more for their practical interest than for how investigation of them may contribute to fundamental research. Furthermore, since grasping the underlying order of these domains requires the grasp of more fundamental laws, interest in investigating them may support to some extent also the interest in fundamental research. Pressures to compromise on (2) and to settle on a qualified version of it as a value of scientific practice derive in large measure from the need for funds to conduct research, for it is rare that funding is provided without strings attached, and without it being open to cuts for reasons that have little to do with scientific schedules and interests. No doubt some sources of funding (for example, universities) may permit a closer approximation to A' than others (for example, corporations) do. In general, funders would be expected to desire that research be conducted with the aim of generating theories acceptable in accordance with I' that could be applied to serve their particular interests well. Thus, the more compromises concerning (2) are entered into, the less likely it is that N', particularly $N'(3)$, would be manifested, and thus the strong manifestation of $A'(1)$ is called into question.

As indicated, gaining theories that accord with I' is not *per se* threatened by entering into the mentioned compromises. Neither is it threatened by the fact, which is consistent with A', that individual scientists, when deciding which theories to entertain provisionally and to adhere to, may make different choices (or follow different hunches) – in ways that reflect their personal or institutional values – about how to further (1). But it may become threatened (at least in some fields of science) where there are strong identities of interests or values (apart from the interest to act according to (1)) among members of the scientific community, scientific institutions and the relevant outside factors. Then the shared values may lead to constraints being put (but perhaps not noticed) on the theories that are provisionally entertained in the scientific community, and thus the exclusion from entertainment of certain kinds of theories simply because they do not fit these constraints rather than because they cannot stand up to empirical test. This suggests that I' might be more constantly manifested if item (e) of $A(2)$ were replaced or, preferably, complemented by:

(e$'$) ensures that widely diverse value complexes are held among the members of the scientific community.

This would also be conducive for movement towards the fuller manifestation of $N'(3)$, and for countering the residue of values that may be left (unnoticed) from the provisional entertaining of and early adherence to theories.

The *autonomy* of the processes and practices of science is a value that is often articulated by the scientific community, though in the hands of the spokespeople of scientific institutions it may be a compromised version that is intended. A' expresses a view about how scientific practices are (ought to be) cognitively self-contained and self-governed by the community of its practitioners. It includes (3), a view about the "autonomy" of scientific institutions – free from outside interference while provided with appropriate resources. Now, an institution's values need not (fully) coincide with those expressed in the practices whose conduct it supposedly enables (MacIntyre 1981: Chapter 14), and in (for example) a corporation's research unit the direction of research may have little to do with the general interests of the scientific community. Nevertheless, it is not uncommon for references to the *autonomy* of science to emphasize (3) (or not to clearly distinguish (2) and (3)), perhaps because it is assumed that *institutional autonomy* is a prerequisite for the high manifestation of (1). That, of course, is a matter for historical and sociological investigation.

Clearly A' is a value of relevance to the interaction between science and the general public (as well as government, business and other powers). Why should the public endorse it? Particularly, why should it not only abstain from interference with the internal workings of science, but also provide it with the positive conditions needed for its pursuit? Why, for that matter, should the public accept that A'(1) is furthered more fully by a community, institutionalized as in (3), behaving as in (2) rather than in the course of a broad based dialogue with groups representing a variety of social, cultural and moral interests? Since the products and effects – cognitive, technological, ideological (for example, Lewontin 1993; Longino 1990; Harding 1993) – of scientific institutions have major impact on the public, the public's interest in the autonomy of science is tied to "autonomous" scientific practices serving its more general interests. As a recipient of public support and a significant source of the conditions that frame and transform contemporary lives, science's claim to *autonomy* thus has to accompanied by a responsibility to the public.[14]

Item (f) of (2) holds that the responsibility of the scientific community to the public is appropriately and adequately exercised in effect by following its own methods, and consistently and constantly producing theories whose acceptance manifests both *impartiality* and *neutrality*. Now, I think it is true that it is in the public interest to support scientific practice insofar as its accepted theories manifest *impartiality* and *neutrality* – both of them; and thus to endorse A', provided that there is reason to believe that the pursuit of high manifestation of (2) furthers that of (1). The other side of responsibility is accountability. *Autonomy* does not excuse science from accountability to the public. Part of that accountability, I suggest, involves empirical (historical and sociological) investigation of such questions as: Does scientific practice conducted in accordance with (2) by an "autonomous" community in "autonomous" institutions lead to theory choices that are *impartial* and *neutral*? Do these choices serve the public well, or do they serve principally certain special interests? Allowing for possible variation from field to field, does item (e) or (e/e') of (2) better contribute to generating *neutral* products?

It does not serve public interest for outside factors to use their influence or power to undermine that theories be accepted in accordance with I'. It may, however, depending on the outcomes of the empirical investigations just mentioned, serve public interest to have (e) supplemented by (e') in (2), if there were reason to hold that *impartiality* and *neutrality* would thereby be enhanced. Since *autonomy* is subordinate to *them*, the responsibility of the scientific community would require that it make the appropriate adjustment in (e). If it did not, there would be no objection, deriving from the priority of *impartiality* and *neutrality*, to the public withholding its support for research pending the adjustments being made. Public intervention into scientific practices and the institutions in which they are conducted, of course, is full of risks, for the line between matters pertaining to item (a) and those pertaining to (e, e') in A(2) can be a fine one in practice. It is not called for to the extent that the scientific community is exercising its responsibility adequately.[15]

The public, then, may have an interest in a variety of value perspectives being represented among the members of the scientific community because of its interest in *impartiality* and *neutrality* – and also for reasons of social justice to open up social spaces that have largely been closed to members of certain groups. Either way, it seems appropriate to extend the domain of acceptable compromise to include matters pertaining to (e/e') as well as (a) and (d). I do not suggest, however, that (e/e') be granted general priority over (c), but that the scientific community be restructured over time so that the two become compatible and mutually supporting.

The above discussion has been premised on the assumption that *impartiality* and *neutrality* are adequately expressed in I' and N' respectively. The issues just discussed, especially that of the membership of the scientific community, become more complicated and pointed when this assumption is rejected. Then *autonomy* will have to be reconsidered (Chapter 10).

The central objectives of this book are to produce elucidations of the view that the sciences are value free and of its three component ideas, and to assess exactly what is and is not defensible in them. We will see (Chapter 10) that the provisional theses stated in this chapter, in view of the arguments accumulated in the intervening chapters, cannot be sustained, but revised versions of *impartiality* and *neutrality* in different ways (though not *autonomy*) can be defended.

5 Scientific understanding

Impartiality is the "rock-bottom" component of the idea that the sciences are value free. *Neutrality* and *autonomy* both presuppose it. It, in turn, requires that the cognitive values be distinct and distinguishable from other kinds of values. I have not yet addressed how to identify cognitive values and deal with controversy about their identification. To remedy this lack I will draw upon an idea of "scientific understanding." I have already referred to the cognitive values as "criteria of cognitive value" (Chapter 3); now I will portray them as "indicators of sound scientific understanding."

The provisional statement of *impartiality* (I') presupposes that theories fit the constraints of the materialist strategies. Is fitting these constraints itself an indicator of having gained sound scientific understanding? Is it a cognitive value? If not, what is its ground? If it involves the play of (non-cognitive) values, does its requiring fit with the materialist strategies contradict the claim of I' that values play no role in the accepting and rejecting of theories?

Modern science is marked in fact by the almost exclusive adoption of materialist strategies in its investigations. I will argue that this rests on mutually reinforcing interactions between the materialist strategies and what I will call the "modern values of control." I introduce the argument in this chapter and subsequently develop it in detail (Chapter 6 and parts of Chapter 7). At the same time, I will argue that adopting the materialist strategies defines only *one*, among in principle *many* approaches – two examples of which will be sketched in Chapters 8 and 9 – to gaining scientific understanding, in all of which the cognitive values ground the sound acceptance of theories.[1]

IDENTIFYING COGNITIVE VALUES

The grammar and logic of cognitive values parallels in many ways that of other kinds of values (Chapter 3). A cognitive value appears in various modes: (including) manifested in theories, partly constitutive of the scientific theory-choosing practices in which it is expressed, articulated in words, and embodied in social institutions. Thus, there can be discrepancies between the cognitive values that are manifested and those articulated. Adopting a set of cognitive values implies a commitment not only to reduce and (if possible) eliminate such discrepancies, but also the capability to defend that it is the "right" set; and thus, paralleling the adoption of values in general (Chapter 2), to defend that manifesting these cognitive values is possible, and that *ceteris paribus* they are indicators of cognitive value – indicators of having gained scientific understanding.

Disagreements about cognitive values

Removing the discrepancies can involve modification of the cognitive values either as manifested (and thus changes in theory-choosing practices) or as articulated. Those expressed in theory-choosing practices, and thence manifested in the theories that are chosen, may need to be modified in the light of argument at the level of articulation; and sometimes those articulated will need to be modified in view of reflection upon actual theory-choosing practices and the possibilities they admit for the manifestation of supposed cognitive values. By way of this two-way interaction between the modes of manifestation and articulation, cognitive values become "reflectively endorsed" (Anderson 1995a).[2]

Modifying one's own adopted cognitive values may be furthered or hindered by which ones are embodied in social institutions. A fundamental shift of cognitive values within a tradition of inquiry (as occurred, for example, in the scientific revolution of the seventeenth century) requires an accompanying fundamental transformation of the institutions that support the inquiry. Personal adjusting of cognitive values will always be constrained in the light of those that are actually embodied, precisely because one's cognitive values partially constitute theory-choosing practices that one shares with some community. So we would expect that the cognitive values that are adopted will be susceptible to variation with the theory-choosing practices to which one is exposed in virtue of the historical moment, culture, class and personal background. This provides the point of entry for some of the radical arguments, which broadly fall under "social constructionism," that in reality cognitive values are subordinate to moral and social values, and that institutional inertia and power are the principal causal factors for the scientific beliefs that we hold.

Given the bipolar sources of the modification of cognitive values – that people (and communities) remove the discrepancies between the manifested and articulated systematically in different ways and thus actually adopt different sets of cognitive values – is neither surprising nor *per se* a threat to rationality. Personal (and group) judgment, and thus a measure of contestation, cannot be eliminated from the use of value words. In coming to adopt cognitive values, however, it seems wise to seek out the sources of disagreements, to take alternative traditions seriously, and even to entertain the possibility that the social conditions for the resolution of disagreements may not presently exist (MacIntyre 1988). Encouraging critical confrontation between articulately developed alternatives, as part of the process of coming to adopt cognitive values, is not a denial of rationality; nor is exploring anticipatory alternatives or attempting to reclaim the contributions of cultures that have been suppressed. On the other hand, rationality is denied when it is held that theories, and the practices in which they are gained, are not subject to critical discussion; when theory-choosing practices no longer are profiled against the horizon of the ideal of truth.

This point has become clouded in some recent discussions, connected with multiculturalism and diversity, where it has been claimed that any attempt to submit cognitive values to critical discussion – to show how they are partly constitutive of rationality – is really an attempt to impose a particular set of cognitive values that is reflective of the interests of a particular culture or ideology. Ironically, reasons are usually given for such claims. Yet an important point is being hidden by this paradoxical rhetoric; it is a critique of the authority and cognitive (epistemic) privilege that has been granted to modern science and, to a lesser extent, certain modes of philosophical, historical and political discourse.[3] Science (as conducted under the materialist strategies) has been widely taken to provide the exemplary expression of rationality, so that the cognitive values thought to be expressed in its theory-choosing practices, under the interpretations they gain in them, have become identified with rationality. It is not paradoxical to challenge rationality *so conceived* and its deep and prioritized embodiment in institutions committed to the gaining of knowledge, for example, to challenge that rationality so conceived in implicated in the false identification of "fitting with particular strategies" (for example, materialist) as a cognitive value, which then serves to disguise a link between this conception of rationality and specific values.[4]

Cognitive values are constitutive of "acceptable" scientific theories. It is in virtue of their manifestation that, according to I', we (should) make theoretical choices. In order to be identified as a cognitive value, I suggest that an attribute of theories should meet two conditions:

1 It be needed to explain (perhaps under idealization or rational reconstruction) theory choices that are actually made, and the character of controversies engaged in by the community of scientists.
2 That it is a criterion of cognitive value – an indicator of sound scientific understanding – be well defended.

McMullin (1983; but cf. 1993) proposes that only the first condition need be met. Commenting on the following passage from Kuhn: "Though the experience of scientists provides no philosophical justification for the values they deploy (such justification would solve the problem of induction), those values are in part learned from that experience and they evolve with it" (Kuhn 1977: 335), he says: "This is to take the Hume-Popper challenge to induction far too seriously. ... The characteristic values guiding theory-choice are firmly rooted in the complex learning experience which is the history of science; this is their primary justification, and it is an adequate one." (McMullin 1983:19) There is no good reason to hold, however, that the values guiding theory choice that are rooted in this complex learning experience are necessarily all cognitive values. The common practices of the tradition, at least in particular epochs, may lead to non-cognitive values also playing a role in theory choice in ways that may not readily be recognized, perhaps because of widely shared metaphysical or value assumptions within the scientific community (and within the social institutions that support its work). I argue on p.107 that fitting the materialist strategies is an instance of such a value. The second condition is needed to separate values like it from the cognitive values.

Cognitive values must carry both explanatory and normative burdens. They play their roles in a context that not only makes genuine contact with scientific practice, but also recognizes that scientific practice is open and, in the long run, responsive to rational criticism.

Criteria actually used in making theory choices

In drawing up the list of cognitive values, then, the first task is interpretive: to reconstruct key episodes of theory choice and controversy in order to identify the criteria that can reasonably be held to have been deployed by the participants in these episodes. In doing so, relevant matters to take into account *include*:

- The criteria that scientists, who innovate, make decisive choices or engage in important controversy, claim to be using.
- Appraisals of how well the proclaimed criteria explain the choices to accept theories actually made, and of whether there are gaps between what the scientists claim and how they act in practice (Laudan 1984); and if there are gaps, plausible proposals about how to fill them.
- The criteria appealed to (for example, in textbooks and critical reviews) in support of a theory (of certain domains of phenomena) becoming held to be part of the stock of knowledge.
- The assent of working scientists to proposed criteria of theory choice.
- Variations of criteria across fields, episodes and epochs, and the reasons given for the variation.

In short, the relevant reconstructions should be grounded in detailed interpretive historical and sociological studies, and interaction with the critical reflections of working scientists. The list of cognitive values (Chapter 3) is generally well grounded in this way, since it draws heavily on Kuhn (1970, 1977) and McMullin (1983, 1993, 1996) who derive their lists from detailed interpretive historical studies; and thus it generally has the ring of plausibility. Even so, as a list of the criteria *actually used* in theory choice (in many fields of science), I think that it is incomplete – in a way that makes a difference.

In particular, empirical adequacy seems to function in interaction with another criterion that helps to interpret it. Empirical adequacy concerns "the quality of fit between theory and empirical data." This "quality of fit" is enhanced by restrictions upon both theory and data that seem to be taken for granted, restrictions that may be construed as showing that another criterion is being brought into play in reaching a balance of the cognitive values. In Chapter 4, I assumed that such restrictions were in place and that they derived from the materialist strategies. I did not include them in the statement of "empirical adequacy" (Chapter 3), so that I could highlight that fitting the materialist strategies serves as a further criterion for accepting theories, or at least for rejecting them (since they permit only theories that meet the constraints to become candidates for acceptance). That fitting the materialist strategies functions as such a criterion leaves it open whether it constitutes an additional cognitive value, or whether it gains this function from some other source, such as metaphysical beliefs or shared social values.

Which criteria are cognitive values?

I turn now to the second task proposed on p. 91. How do we determine that a criterion actually (or said to be) used in choosing theories is, or is not, one of the cognitive values? How can we determine whether or not it is a criterion of a theory's cognitive value?

Broadly speaking, there seem to be four relevant kinds of considerations that have been entertained:

From general theories of knowledge.
From evolutionary naturalist and cognitive psychological theories of knowledge acquisition and appraisal.
From arguments about the possibility, or impossibility, of the concrete manifestation of the proposed criterion in a theory.
From whether or not it serves the objectives of science.[5]

Considerations of the first kind have sometimes sustained attempts to ground rule-governed accounts of scientific rationality; they have also been used in the context of defending realist interpretations of science (McMullin 1993). Those of the second kind may build upon accounts of human cognitive faculties (Ellis 1990), or come to propose that scientific rationality is fundamentally social (Solomon 1992, 1994). Those of the third kind account for the absence of certain items from the list, for example, "certainty," either of the Aristotelian kind ("necessity" or "intuitive") or the Cartesian ("a priori"), since the character of our scientific practices does not permit us to anticipate (or even to recognize) the concrete manifestation of such a value. I (without wishing to minimize the importance of the other considerations) will concentrate exclusively on considerations about the objectives of science.

Here complications abound (Laudan 1984). It is a difficult interpretive task to discern objectives, and there is pretty intractable disagreement about them. Moreover, we cannot discern them without attempting to explain why we use the criteria we actually use (and to identify which ones we should use) in choosing theories; so that identification of objectives and picking out which criteria are cognitive values are deeply intertwined tasks. Moreover, objectives may vary with field, epoch and even school of thought.

While recognizing this, let us consider, as a candidate for being an (the) objective of science, the following:

O₁ The objective of science is to represent phenomena (in rationally acceptable theories) in terms of their being generated from underlying structure, process and law, and thence to discover novel phenomena.

Something like O_1 has often been considered the objective of science within the tradition of modern science, and it is rooted in the influential Galilean (metaphysical) idea (Chapter 1). Given it, a case can readily be made (though I will only hint at it here) that most (not all) of the items on the list (Chapter 3) indeed are cognitive values, characteristics of theories whose acceptance serves O_1. Explanatory power, for example, is such a characteristic – provided that an explanatory ideal is assumed: to explain is to represent phenomena as generated from underlying law, process and structure. O_1 seems to carry with it an explanatory ideal: explanation is materialist and (often reductionist), but it is not, for example, teleological and it need not be deterministic. Choosing theories informed by the items on the list (perhaps, as in the case of explanation, under particular interpretations) serves O_1.[6]

Following a similar argument, adopting the materialist strategies also serves O_1. Then, fitting the materialist strategies would appear to be an additional cognitive value, one perhaps that significantly affects the interpretation of all the other, particularly "empirical adequacy." O_1, thus, seems to vindicate that I' provides a suitable statement of *impartiality*. Under the materialist strategies, scientific theories represent phenomena (objects, entities, beings, things, events, fields) simply in terms of their hypothesized structures, processes and components interacting with one another in a way that can be represented in mathematically-formulable laws – abstracting the objects from any value they may be bearers of, or any place they may have in human practices or social structures. From pursuing O_1, nothing follows directly about the relevance of theories, and of the phenomena discovered in the course of scientific practice, to human practices in general and to the objects of ordinary experience.

Why do we attempt to gain understanding of natural objects through cognitive practices that abstract them from the contexts of human practices in general, from their role in ordinary experience and from the possibilities open to them in these contexts? (Why are O_1 and its variants often taken to articulate the objective of science?) To address this usefully, we need to consider generally what it is to gain understanding. This will open up space for proposing various kinds of objectives that might be entertained for science. It will also provide the context for addressing sharply the question of whether fitting the materialist strategies represents a cognitive value.

UNDERSTANDING

What are natural objects and phenomena, and what is it to understand them? The answers vary with context: focus of interest, practice being engaged in, background knowledge, and participants in the discourse. Regardless of context, however, understanding things (events, states of affairs, phenomena) involves the following interacting components:

- An account of *what* they are: of the kind of thing they are; of their properties, behavior, relations and their variations with time.
- An account of *why* they are the way they are: why they have the properties, behavior and relations that they do; why they have varied in the way in which they have; an account of their origin.
- An account of the *possibilities* (including hitherto unrealized ones) that they allow in virtue of their own powers to develop and by means of their interactions with other things.

Understanding reality involves grasping the "what?," the "why?" and the "what possible?" of phenomena.

Each of the three components is open to an array of interpretations. Regarding "what?", an object may be considered as: an object of experience; an object of a human practice, something acted upon or interacted with in practical life for the sake of personal, social or institutional ends; and an object which manifests causal relations with other objects – whether in virtue of lawful connections among the phenomena in which it participates and others, or of its place in a structure or a system (for example, ecological, social), or (in the case of human beings) of intentional and communication relations.

Clearly, depending (contextually) on what an object is considered to be, the answers to the "why?" and "what possible?" questions will take different forms and reflect different interests. To illustrate, consider the seed (for example, of wheat) and ask what possibilities are open to it (Chapter 8). The seed can be considered to be many things, of which I will pick out two: an object which generates, on cultivation, crops with quantifiable yields; and an object which is integrally a part of social processes. Considered the first way, seeds may become hybrids and genetically engineered such that, when cultivated under specific conditions, crop yields increase significantly. Considered the second way, in becoming the generator of such higher yields, the seed also becomes a commodity – something produced and cultivated in capital-intensive enterprises and bought and sold on the market – rather than an object produced annually as a routine part of the crop. It thus becomes a different object of human practice related differently to the social order. What is a quantitative enhancement when the seed is considered one way is identical with a fundamental social change when considered another way.

Understanding, then, can take different forms and, in so doing, be responsive to the different interests of different practices. Understanding sought under the materialist strategies is just one form. Why it should be granted primary emphasis needs explanation. I turn now to elaborate this brief introductory statement, and to give a systematic characterization of a number of different forms of understanding.

Forms of understanding: wide-ranging and full

What is it to understand a phenomenon (event, state of affairs, thing, object in the broadest sense: including material object, person, social institution)? What do we attempt to identify when we aim to understand an object? As indicated on p. 95, what we identify varies with the particular context or focus of the quest for understanding, and so with the particular way in which the question is put. Understanding an object always involves being able to explain it and to identify the possibilities open to it. It also involves meeting additional cognitive (epistemic) conditions. Understanding is offered systematically, for example, in (or with the aid of) theories, or in narratives, that manifest the appropriate cognitive values to the appropriate degree.

Much of the argument of this book deploys (explicitly or implicitly) a distinction between wide-ranging and full understanding. In order to introduce it, I offer first a list of kinds of items that may appropriately be offered (in some contexts or other) as components of *understanding an object*, or *kind of object*, recognizing that some amount of provisional classification (some provisional answer to "what") is a prerequisite to gaining understanding.

When aiming to produce understanding of objects we identify (posit) some of the items on the following list:

1   Their components, how they are structured, and the kind of processes they may be part of.
2   Their properties and relations, present and past.
3   The principles with reference to which their movements, variations and interactions can be explained; and particular conditions whose causal roles are represented in the principles.
4   The possibilities open to them to:

a   change (to exhibit different properties and relations; to develop into);

b   affect as causal agents;

c   become upon decomposition; and

d   become upon becoming constituents of other objects, environments, practices or systems;

and the conditions under which these possibilities are realized.

5   The conditions which brought them into being, and (where applicable) those which sustain their existence.

6   The other objects or systems with which they share the same explanatory principles, and the particular conditions (often instances of variables of the principles) that account for different variations in the different systems.

7   Their relationships with the environments – physical, ecological, human, social, (sometimes) spiritual – in which they are located, and the sort of reciprocal interactions they exhibit with one another, including how they relate to us (human beings) and what we can do with them, and so their places:

a   in the realm of daily life and experience;

b   in various human practices, and in the institutions in which they are conducted;

c   specifically in the practices of investigation: their interactions within these practices and the similarities and differences with their interactions in other contexts – paying attention to the material and social conditions of the investigatory practices and the institutions that provide them, the broader human practices that can be informed by the outcomes of the investigatory practices (and their human, social and ecological consequences), and identifying what can be done with them following investigation.

Not only does gaining a comprehensive form of understanding, that systematically offers items of all these kinds, seem to lie beyond our powers, but also the items are open to various interpretations and different emphases. It is this that makes possible that there be different forms of understanding. Think of "principle" in item (3). Principles can have different characters: lawful, intentional, teleological, functional or specifying location in relation to some whole. Not all can be simultaneously in play. Depending on the character of the principles, different items will have greater or less significance, and sometimes different items might be in tension. Yet probably there is no form of understanding in which there is not at least a hint of every item. I will identify some important forms of understanding.

Wide-ranging understanding

The understanding, gained under the materialist strategies, is an instance of what I call "wide-ranging understanding," a form of understanding that gives special emphasis to item (6). Ready responsiveness to this item is most available when principles are interpreted as laws, which express (possible) relations among quantities. Then, it highlights quantities in (2), and – in order to be able to represent the lawfulness of phenomena more completely – it may offer posits about the underlying components of the objects (characterized quantitatively), about how they are structured and about the processes into which they enter and the laws they reflect. It aims to consolidate principles which, when deployed in conjunction with posits about underlying components, structures and processes, produce understanding across the widest range of experimental, technological and natural spaces. The same principles play the key explanatory roles across all of these spaces, where the different phenomena in the different spaces are accounted for in terms of variation of the values (boundary and initial conditions characterized quantitatively) of variables of the principles (Taylor 1970).

These principles, together with the accompanying posits about the underlying order, encapsulate the material possibilities of these spaces, those that can be identified in terms of being generated from underlying law, structure and process, given the appropriate boundary and initial conditions. The material possibilities do not exhaust the possibilities of things in the spaces. In the case of the seed, the increased crop yields are among the material possibilities encapsulated in the relevant biological theories, but not the seed becoming a commodity. The latter possibility cannot be identified apart from the context of social relations. Also the obtaining of the boundary and initial conditions typically cannot be explained without reference to this context. Wide-ranging understanding abstracts from the human, social and ecological characterizations that also fit these spaces; it largely ignores item (7). It aims to be context-free understanding. Abstraction from context and wide-rangingness of application go hand in hand. Understanding can be more or less wide-ranging.

Full understanding

There is another form (more accurately, set of forms) of understanding that grants high salience to (7). I call it "full understanding." It seeks, to the extent possible and deemed worthwhile, to understand objects in all (the fullness) of their dimensions, aspects, concreteness, wholeness and particularity; and, to the extent possible, identify all their significant possibilities, including those that the process of gaining understanding opens up. It does not dissociate gaining understanding of an object from understanding what one is doing – what possibilities of the object (not just material ones) are being identified, and what are their links with the material and social conditions of the process and with the institutions providing them – when engaged in the understanding-gaining process. Full understanding does not ignore (6). In principle, it attends equally to the material possibilities of spaces (and so may be informed by wide-ranging understanding), and to the human and social characterizations of their boundary conditions, to the human, social and ecological consequences of processes within the spaces, and to the human and social possibilities that may be hidden in them.

But identifying (6) and (7) can be in tension. Not all the possibilities of an object can be realized simultaneously; and obviously the realization of some of them may preclude the realization of others. Equally, since investigation requires significant material and social conditions, the investigation of one class of possibilities may effectively preclude that of another. Investigating the material possibilities of the seed may, contextually, preclude investigating the seed's possibilities under social relations where it does not become a commodity; pursuing wide-ranging understanding of the seed may, contextually, conflict with attempting to grasp its place and its possibilities under a desired set of social relations (Chapter 8). Where there is such tension, a balance is to be sought. Priority is not automatically granted to (6); it may be subordinated (Chapters 6, 7 and especially 8). Thus, while full understanding may draw freely from the results of wide-ranging inquiry; it may query the general significance of the drive, that tends to motivate scientific institutions today, to keep expanding wide-ranging understanding (Chapter 10).

Such tensions show that full understanding is necessarily bounded. Often we cannot aspire to identify all the possibilities of the objects of investigation. Then one must choose which class(es) of possibilities to investigate. (Investigating only the material possibilities represents a choice.) The choice will reflect one's place in prevailing social relations and, no doubt, one's values. The ideal behind full understanding pushes for the recognition of multiple possibilities, including "lost possibilities" (Chapter

8), for the multiple-facetness of all objects and of the consequences of their uses and interactions, and for the fact that the achievement of any possibility can always be described with different categories, where the different descriptions will be more or less pertinent to the valuation of the achievement. In the case of the seed, augmenting the crop yield is (contextually) the same achievement as turning the seed into a commodity. While fullness is bounded, full understanding aims for an array of characterizations of objects that is sufficient to enable reflective valuations of the possibilities of objects that become grasped in the course of the investigation.

Scientific (systematic empirical) inquiry

A number of objections have been made to my account of full understanding. It has been said that science is, by definition, conducted under the materialist strategies,[7] so that the discussion of full understanding has little to do with science. It is, however, also of the nature of science to involve systematic empirical inquiry. My point is that the latter need not be conducted exclusively under the materialist strategies. Whether or not we call other efforts to engage in systematic empirical inquiry "science" is of little moment. The question of why are the materialist strategies adopted still remains; it just gets turned into: why engage in "scientific" rather than some other form of systematic empirical inquiry? In order to maintain a reasonably concise terminology, I will treat "scientific inquiry" as equivalent to "systematic empirical inquiry," so that "modern science" (conducted under the materialist strategies) is considered one approach to scientific (systematic empirical) inquiry. Then, I will use "theory" to refer to any "systematic empirical body of posits," so that a body of traditional knowledge about varieties of plants and their ecological relations will be called a theory, as well as a mathematically formulated theory of physics and a narrative of the evolutionary development of *homo sapiens*.

A second objection[8] is that, although objects are multifaceted, we have learned as modern science has developed that the most effective way to investigate them is from the perspectives of a variety of autonomous disciplines, each of which investigates the possibilities of objects from its own particular perspective, so that, for example, in the case of the seed it is entirely proper to separate the possibilities of higher yields and the economic and sociological possibilities. While there is a large measure of truth in this point, it ignores the tension between identifying (6) and (7) discussed on p.99: the social conditions for realizing the possibilities for higher yields from the hybrid varieties may eliminate the social conditions in which the seed might not become a commodity. It may not be possible to investigate the "two" possibilities separately. In the concrete context of the green revolution, for example, the "two" possibilities may be realized in one and the same achievement. Then it would not be appropriate or desirable to attempt to demarcate the material possibilities from the full range of possibilities; rather the explanation of the achievement should reflect the inseparability (in the specified social context) of the economic and the materialist causal factors.[9]

A related objection is that I have conflated the distinction between fundamental and applied research, and that full understanding is appropriate only in the context of applied research. It is true that applied research is unintelligible outside of its social and ecological locatedness. However, when conducted under the materialist strategies it tends not to deal adequately with the multifacetness of things, and in particular to the contextual incompatibility of realizing some different classes of possibilities, and thus to the issue of "lost possibilities" (Chapter 8). It asks: How can we realize the material possibilities of things that we have discovered in useful ways (in general or in service of a specified goal) without also generating undesirable side effects? The side effects considered generally are those that themselves can be understood as generated from underlying law, structure and process; others tend to be considered in an *ad hoc* way. It does not tend to investigate the effects of applications[10] on prevailing social arrangements (though in recent years there has been much pressure to investigate "environmental impact"), and to ask, for example: what possibilities do we need to identify in order to strengthen a certain social arrangement, to preserve an ecological system, and to avoid undesirable consequences for human lives (Tiles 1987)? Asking questions of these kinds we presuppose neither that there is a sharp separation between fundamental and applied research, nor that wide-ranging understanding (alone) will be able to play a useful role.

Full understanding abstracts the science neither from the sociology and the ecology, nor from the practices and institutions that generate the science. Full understanding thus has a critical component as it addresses the dominant scientific practices, and a positive investigative component that is explicitly informed by social values, posing such questions (that the tradition of modern science usually presumes to be settled) as: Is wide-ranging science genuinely context free? And, can the material possibilities of spaces be explored and charted generally when one abstracts from the social arrangements and practices that shape these spaces?

OBJECTIVES OF SCIENCE

I suggested on p. 99 that adopting the materialist strategies makes clear sense if one considers the objective of science to be given by O_1 (to represent phenomena as generated from underlying structure, process and law). Why adopt O_1 as the objective of scientific (systematic empirical) inquiry? Well, one might say: if not O_1, then what? Here is an alternative proposal:

O   The objective of science is to gain understanding of phenomena. This *includes* to encapsulate (reliably in rationally acceptable theories) possibilities that are open to a domain of objects, and to discover means to realize some of the hitherto unrealized possibilities.

O is more encompassing than O_1. So, too, are the following:[11]

O'   The objective of science is to provide (in rationally acceptable theories) a literally true account of what the world is like.
O''   The objective of science to provide (in rationally acceptable theories) the best explanatory account of natural phenomena.

Adopting O_1, I indicated, may be partly motivated by Galilean (materialist) metaphysics. O, O' and O'' may not. They can be interpreted to mean more or less the same thing, but they connote different images. O' suggests that the objective is to gain a detailed, accurate map of the spatio-temporal totality that is the world. O suggests, rather, that the world be considered an inexhaustible well of possibilities to be probed over and over again, but never exhausted. O is motivated by emphasizing that what is the case (what has been, and what is actually realized, and the projection into the future of its regularities and structures) cannot be identified with what can be. There are possibilities that fall outside of currently actualized regularities and structures, so that an empirical charting of what *is*, and of what its regularities *are*, does not suffice to sum up what *can be* or what genuinely could have been. The real is not exhausted by the actual; it includes also the genuinely, as distinct from merely logically or imaginatively, possible (Bhaskar 1975; Lacey 1997c). O'', unlike O', makes clear that explaining as well as describing phenomena lies within the purview of science and it is not tied to any one explanatory ideal. Moreover, being able to explain the current condition of phenomena is a necessary condition for encapsulating reliably the possibilities that may be open to them, including the novel possibilities that may remain hidden when we attend only to current regularities. O contains O''. Both because of its generality and because the links with values (that I will discuss) draw attention to alternative classes of possibilities, I will consider O as the salient alternative to O_1.

O is more general than O_1, but following O_1 is consistent with following O. Following O_1 we do encapsulate possibilities, but only those that can be represented as generable from the underlying order. Following O beyond where O_1 takes us, we are open also to those possibilities that can only be described when we do not abstract objects from their human, social and ecological contexts. When our focus is upon spaces in which human agency is not relevant as a causal factor (Chapters 6 and 10), the success of modern scientific inquiry suggests that O reduces to O_1; so O retains contact with scientific practice that accords with the Galilean idea. Departures from O_1 tend to concern the realm of daily life and experience, and currently to be important principally in the human sciences and in those areas of biology that impinge upon human capabilities and social arrangements (Chapters 8 and 9). One is drawn to adopt O when one keeps in mind that scientific knowledge is there to be applied, accepted for the end of informing practical projects as well as cognitively; and when one asks: What kinds of systematic empirical understanding are needed to inform all of the great variety of human projects?

Different approaches to science

Adopting O permits that there may be systematic empirical inquiries in which relevant data are not selected to fit the materialist strategies. So, under O, fitting the constraints of the materialist strategies will not be considered a cognitive value. Indeed, if we adopt O, we might consider O_1 as the objective of a particular (very important) *approach* to scientific inquiry – the materialist or the Galilean/Baconian approach – rather than the objective of science as such In the next chapter, I will justify defining this approach in terms of a link with social values, the "modern values of control" (Baconian utility) by arguing that O_1 bears a close affinity with an objective that can be put as follows:

O_1' The objective of the Baconian approach to science is to encapsulate (reliably in rationally acceptable theories) the possibilities of a domain of objects that would serve well the interests of the modern values of control, and to discover means to realize some of its hitherto unrealized possibilities.

Parallel to this, there could be defined a class of approaches to scientific (systematic empirical) inquiry, where each approach (O_j) bears a close affinity with an objective which is an instance of the following schema:

O_i' The objective of the (...) approach to science is to encapsulate (reliably) the possibilities of a domain of objects that would serve well the moral/social project (...), and to discover means to realize some of its hitherto unrealized possibilities.

Filling in the " ... " with "grassroots empowerment" we get O_2', linked with an approach (O_2) whose strategies involve links with traditional, local knowledge in impoverished countries (Chapter 8); with "feminist" we get O_3', linked with an approach (O_3) that deploys strategies that constrain theories in the direction of "complexity, ontological heterogeneity, interaction and non-reductionism" (Chapter 9). In general, which strategies are adopted will follow readily from the approach to science adopted.[12]

Not only O_1, but also all of O_i (in principle) may define concrete – perhaps complementary – ways to further O, where in each approach a different class of possibilities is investigated. I am tempted to replace "possibilities" in O with "the full range of possibilities." Then O_1 and O_2, for example, could definitely be seen as complementary. But the pursuit of a particular approach, such as O_1, has material and social conditions, which may effectively preclude the pursuit of another approach, such as O_2. For instance, the exploration of one range of possibilities may effectively, in the light of necessary material and social conditions, preclude the exploration of another (Chapters 7 and 8). The perceived complementarity may only be abstract and not open to effective realization under concrete social circumstances.

Why adopt the materialist strategies?

O is a very general encompassing objective, a critical reference point from which to discern the partiality of approaches that deploy particular strategies, but not a positive source of research guidance. Only the adoption of a strategy provides positive direction to research, but O does not guide us towards any particular strategy. The materialist strategies are widely adopted, and it is often held that they ought to be adopted virtually to the exclusion of other strategies, that somehow the objective O is best furthered by following the approach O_1. Why? Why is research conducted under the materialist strategies considered exemplary? Why is inquiry aiming to encapsulate the material possibilities of things considered exemplary? These questions need to be answered by the scientific community at large and the body of its institutions, and not just by individual scientists.

Historically three kinds of answers have been given to these questions. The *first* appeals to materialist metaphysics: the world really is such that all phenomena are generated from underlying structure, process and law. Thus, a theory that soundly represents the underlying structure, process and law of a domain of phenomena will *ipso facto* represent relevantly its causal structure, and it will suffice to encapsulate its possibilities for they will be exhausted by the material possibilities of the domain (Chapter 6).

A *second* answer is that adopting the materialist strategies serves the interests of Baconian utility: understanding gained from following them, and effectively only them, includes that which enhances the human capability for exercising control over nature. A variant of this would be that it provides the kind of understanding needed to grasp the dominant objects and leading practices of modern practical life and lived experience, and thus to live effectively in the modern world. To account (rationally) for the virtual exclusivity of adopting O_1, we must add that the interests of Baconian control are generally compelling and in conflict with the value complexes implicated in the definition of the other approaches.

The sheer intellectual interest of theories that represent underlying order, combined with the fact that we have a very successful "track record" in establishing theories of this kind, is sufficient for a *third* answer. Theories, gained under these strategies, often manifest the cognitive values to a high degree with respect to a broad array of experimental data, supplemented by the data of successful practical applications and of some natural phenomena. Here, one might say, the stock of knowledge is increasing. Moreover, there are good reasons to hold that following the same strategies will enable the stock of knowledge to continue to grow; and, at the present time, there do not appear to be plausible alternative strategies to explore if we wish to add to the stock of knowledge, especially if we wish to continue to investigate certain kinds of phenomena (for example, astronomical or optical ones). O_1 defines "the only game in town."

This answer may be reinforced by arguments of Putnam (1981, 1990) that the activities and virtues involved in the gaining of scientific knowledge are partly constitutive of human flourishing; that commitment to the cognitive values is "part of our idea of human cognitive flourishing, and hence is part of our idea of total human flourishing, of Eudaimonia" (Putnam 1981: 134). Then, we adopt O_1 for want of alternative ways to pursue O in the institutions actually available for the conduct of inquiry in advanced industrial societies. How powerful the reinforcement is needs to be assessed by balancing Putnam's claim with investigating the contribution of science to "humanitarian" ends (Feyerabend 1979) or to "wisdom" (Maxwell 1984), and in the light of raising the question of whether the kind of human flourishing that Putnam has identified can be achieved in a way that is compatible with furthering social justice. It might be further reinforced by combining Putnam's argument with Kuhn's (1970) analysis that historically an old paradigm has (must have?) a unique successor; if we want to play the "game" of science we would expect that there would be only one "in town" (Chapter 7).[13]

My own view is that the best answer is provided by a suitably refined Baconian one – the second with important input from the third (cf. Tiles 1987). It is difficult to separate the second and the third answers (even if appeal to *neutrality* supports that they are distinct), for satisfying the intellectual interest also tends to satisfy the interests of Baconian utility; and social institutions support the projects of O_1 principally because of their contributions to Baconian utility. Pragmatically, especially in the awareness of the individual researcher, some version of the third answer is usually stated in support of the adoption of O_1. Nevertheless, it is not sufficient. At best it is an argument to consider O_1 worthy of adoption; but it does not provide support for not adopting some other O_i (at least by sectors of the scientific community), even if the practices of that alternative have still to be developed in detail and the institutions that might support them still have to be created. Being the "only game in town" is not a good reason not to attempt to bring other games to town, unless there are good reasons why they should not be brought to town. The absence of an alternative "scientific game" in contemporary universities and research institutions might reflect only that currently hegemonic values have ensured that the necessary material and social conditions for development have been denied (through structurally maintained mechanisms) to alternative approaches that deploy different strategies. The "only game in town" argument is not value neutral if the lack of alternatives is a consequence of denying the conditions necessary for an alternative to develop. That it is not value neutral, as such, does not imply that it is a bad argument; it depends upon whether the relevant values can be defended.

I suggested that, whatever the motives of individual scientists may be, societal institutions provide conditions for scientific research because of its contributions to Baconian utility. It is quite common to say that it is the objective of science to understand "the world we live in" or "the world around us." These phrases tend to be used ambiguously. Sometimes they are intended to refer to the underlying material order and vast material complexity of things in which our lives are located, evoking wonder, puzzlement and perhaps a sense of our insignificance in the overall order of things. But "our world," the world we experience in day-to-day living, has also been shaped historically, so that most of the objects we deal with in daily life and experience bear the mark of human history. Key objects that we encounter in the course of daily life exist (causally) because of their role in implementations of social values, such as Baconian utility and the socio-economic values with which it tends to be co-manifested. Thus, the presence of objects whose behavior is generated from certain specific kinds of underlying law, structure and process – or the underlying order that is pertinent to the general conduct of our lives – will sometimes in part be the outcome (causally) of the adoption of social values; so that the full range of possibilities open to these objects cannot be identified without reference to the social values.

These objects are simultaneously objects of materialist understanding and objects of social value. Any ready extrapolation from the investigation of them to objects of daily life and experience in general runs the risk of imperceptibly importing this social value into other contexts. It can be a good reason to adopt O_1 (at least as a subsidiary objective), that key objects and practices of the world of our lived experience require materialist understanding. It does not follow that values are not crucial in making that adoption, for the possibilities open to the objects are not exhausted by those realizable where these values are predominant, or those that can be predicted under current social constraints. The "world" of daily life and experience is always structured (where the structures have causal implications) in ways that partially represent the embodiment of social values. But the possibilities open to the objects found in this "world" extend beyond those realizable (and perhaps predictable) under the prevailing (or any other) structures. There may be powerful institutional limitations upon investigating possibilities that could not be realized within prevailing structures, precisely because these structures do embody highly certain social values. To confine investigation just to the possibilities realizable within these structures is, therefore, implicated in values (Chapter 8; Lacey 1997c).

I will motivate my view by exploring in the next chapter the relationship that exists between following the materialist strategies (adopting O_1) and the distinctively modern attitude towards the control of nature.

Does adopting the materialist strategies represent a cognitive value?

Where does this leave the question of whether or not fitting the materialist strategies represents a cognitive value? Certainly satisfying the constraints of the strategies commonly functions as a criterion of theory choice (typically as a ground for rejecting certain theories) and, if O_1 were considered identical to O_1, it ought to. One way to regard it may not be as an additional cognitive value (since it cannot be grounded in O by itself), but as a "constitutive value" of the approach O_1. This suggestion is reinforced by observing that the materialist strategies operate as a criterion of theory choice at a different level from those items that I have listed as cognitive values. *Antecedently*, it constrains the class of acceptable theories and selects the class of relevant data; *then* the cognitive values play their role in choosing among the theoretical candidates. As such a constitutive value, it may be considered grounded either in the general features of the chosen object of interest for science (underlying structure, process and law), or in the interest of Baconian utility to explore only the material possibilities of things.

Looked at in this way, the materialist strategies function as a criterion of theory choice in virtue of an interest in the underlying law, process and structure of things[14] and in their material possibilities. They define the class of theories and the classes of possibilities of interest. That interest, and the social values that nourish it, has no implications regarding the specific theoretical posits investigated and confirmed, and regarding the specific material possibilities that are encapsulated. *After* the play of the strategies, the play of the cognitive values sorts out such specifics in the light of the data that are selected and obtained. Indeed, values could have no implications in regard to these matters. Neither can values even ensure that adopting a strategy will lead to success in generating theories that manifest the cognitive values highly. No matter how much interest one might have in encapsulating a kind of possibilities, there is – prior to the outcome of the research – no assurance that there are any such possibilities to be encapsulated. Thus, the great success of generating theories under the materialist strategies that manifest the cognitive values highly vindicates – after the fact – that adopting these strategies indeed is partly constitutive of an approach that contributes to realizing O; but the objective of realizing O is not served uniquely by following O_1 – and thus I' cannot stand as a satisfactory explication of *impartiality*. Different values could lead to interest in different (more encompassing or intersecting) classes of possibilities, the investigation of which might require strategies other that the materialist ones. As long as the role of values in theory choice is played out in the context of the definition of the classes of possibilities of interest (and thence at the level of choice of strategies) and does not extend to the choice of specific theories, nothing paradoxical or logically problematic need arise.

Nothing follows a priori from reflection on the objective O regarding which strategies might contribute to the further realization of it. Anything, in principle, might be tried, for only after the fact, after having attempted investigation under a strategy, does it become apparent whether adopting that strategy serves to further O. It is a condition upon a strategy that can be partly constitutive of an approach to furthering O that it actually produce theories that manifest the cognitive values highly. Thus, for example, we no longer attempt to follow the strategies of Aristotelian science (Chapter 7).[15]

Strategies as constitutive values of approaches *to* inquiry

The cognitive values are the features desired of theories in virtue of their being generated for the sake of furthering the objective of scientific (systematic empirical) inquiry, O. There remains dispute about what the exact list of cognitive values is. I highlight: empirical adequacy, explanatory power (both wide-ranging and full), power to encapsulate possibilities, internal consistency, consonance, source of interpretive power, and rejection of *ad hoc* features. Agreement on the list is not important to the argument that follows for the remainder of the book, but that fitting a particular strategy is not a cognitive value is central. Fitting a particular strategy, however, is a constitutive value of an approach to furthering O, which typically plays its role (logically) prior to the roles of the cognitive values being played. The cognitive values involve an assortment of desiderata for theories, not all of which may be able to be manifested highly in the same theories. Often, for example, there will be a "trade-off" between wide-ranging and full explanatory power. How the cognitive values are to be ranked may vary with the strategy adopted, as will also the interpretation of empirical adequacy, since its meaning remains unspecified until a class of empirical data has been selected (Chapter 10).

The cognitive values do not suffice to pick out any one approach, or – prior to the conduct of inquiry – to identify those approaches that can contribute to furthering O. Every scientific (systematic empirical) inquiry is conducted within a particular approach, partly constituted by the adoption of particular strategies. I will argue that, generally, settling on a particular strategy is linked with its mutually reinforcing interactions with particular social values, and that the conditions of realizability of the possibilities identified under the strategy include social (institutional) structures that embody these values. This point can be obscured where the social values in question seem to be "obvious," or (as in the case of Baconian utility) are part of the deep self-understanding of a culture, for then the high manifestation of these values may readily be taken as a natural (universal) rather than as a historical (culturally specific) phenomenon. The obscuring can be even greater when there are no ongoing counter research programs that offer concrete results challenging such alleged universality. In the next chapter I will argue that adopting the materialist strategies (following O_1)

involves mutually reinforcing interactions with the modern values of control. Later (Chapters 8 and 9), I will illustrate other strategies and the social values with which they enter into mutually reinforcing interactions.

6   The control of nature

While it is part of human nature to exercise control over natural objects, in modern times exercising control has gained distinctive features (Leiss 1972): its extent, its pre-eminence and its centrality in our lives, the high and virtually unsubordinated value granted to it, the dissociation of considerations of control from those of the meaning and value of our activities and social arrangements, the intense efforts to expand and implement our capabilities of control, and the conviction that these efforts will be at the heart of projects to meet human needs and wants even as their embodiments continually generate new needs and wants. Consequently, certain values connected with the control of nature rank especially highly in modern value complexes. I will argue, as anticipated in the previous chapter, that the nearly unanimous adoption of materialist strategies in modern scientific practices becomes intelligible largely in virtue of its mutually reinforcing interaction with these values.

CONTROL

By nature, human beings are reflective, "embodied and active in the world" (Taylor 1982: 101); they are also social and cultural beings. They are agents whose interactions with material objects and other human beings require intentional explanation, in which action is portrayed as following from an agent's beliefs and objectives, and thus understanding. We exercise control over objects when we deliberately and successfully, informed by our beliefs about them, submit them to our power and use them as means to our own ends. Not every effective intentional interaction with the world is an instance of control. Perhaps in some cultures few of them are, but in all cultures under some conditions control is practiced and valued. Control is contrasted with such stances as reciprocity, mutuality and respect, where the value of the object interacted with reflects a measure of integrity accorded to it, and is not reducible to instrumental value for the agent.

The exercise of control is obviously served by practical understanding, understanding of the effects of our actions on things and their effects on us. There exists systematic empirical practical understanding in all cultures, as well as the ongoing interplay between interaction with material objects and practical understanding. Successful interaction provides the decisive tests of this understanding, whose forms may vary, reflecting different ways in which interaction with material objects may be related to interactions with other beings.

Interaction with nature may be circumscribed by its fit with a social, ecological or cosmic order, and by a particular conception of human flourishing. It may, for example, in a given culture, take place within natural rhythms, with a limited set of ends and means defined by traditional practice, where the tradition ensures that, save for unexpected circumstances, there is harmonious and reciprocal interaction with nature. Then, human control of the natural environment is balanced by nourishment and maintenance of it, so that human relationships with and within it can be permanent, and the preservation of the environment sets bounds to acceptable ends. Such a constancy, punctuated only rarely by periodic or occasional variations, can provide the basis for a stable social order where there is deep interlocking of social and cosmic visions. Where this is the case, the form of practical understanding will reflect how interaction with nature contributes positively or negatively to the desired order, and it will explore the possibilities of nature in relationship to those that this order allows. It will be constrained to grasp things with categories related to the social, ecological and cosmic order, and to serving the particular ideal of human flourishing pursued in this order. Where such conceptions and practices are present, the distinctive human stance towards nature is well captured with notions such as attunement, adaptation, harmony, and participation. Control is subordinate to these relations, and limited in scope, valued only to the extent that it contributes to ends circumscribed by the desired social order and ideal of human flourishing. Exploring the possibilities of control beyond these bounds has no moral (or rational) intelligibility. Important echoes of these conceptions are to be found in certain ecological, feminist and third world popular movements today (Chapters 8 and 9).

The place of control in modern value complexes

The distinctive modern attitude towards exercising control over things rejects in general the subordination of control to other stances towards nature and to particular social values and ideals of human flourishing. It looks for the regular expansion of the scope of effective controls throughout the activities of practical life. That expansion has occurred so successfully that practical life has become shaped pre-eminently in the course of implementing novel and far-reaching possibilities of control, principally by way of technological advances. Whether they are connected with energy, transportation, medicine, agriculture, communications, or education, practical problems and social questions increasingly are becoming considered to be properly open primarily to technological address. Consequently the realm of daily life and experience has become dominated by the products of our exercising control over things, and its social institutions transformed and adapted to accommodate and serve the resulting forces, needs, wants and interests of practical life.

Although an historical fact of major importance, the enormous expansion throughout modernity of the successful control of natural things was not inevitable; and, despite the current "realities" of the globalization of the market, reasonable aspirations for the future need not be limited to those that depend upon its continued expansion (though they must contend with the social forces that engender it; Lacey 1997c). Gaining control over material things has become a very highly rated social value – not subordinated in a general way to any other values, but also not unambiguously granted ascendancy over them. Where its interests and those of other prominent social values clash, there is no systematic way to allocate priority or to define compromises. Sometimes control is taken to be a value for its own sake, a power whose exercise is the exemplary expression of human rationality. More commonly and more defensibly, *via* technological advances, it is taken to be able to serve all sustainable social values and ideals of human flourishing, and to serve to enhance human well-being in general and in the long run.

Thus exercising control over things has come to be considered largely in abstraction from links with other values, and efforts to further its expression can proceed with relative autonomy so that matters of viable and desirable social arrangements, and meaningful cosmic order, have tended to become subordinate to the value of control. Not every instance of exercising control is valued, both because human well-being remains a relevant evaluative standard and because in some instances it may conflict with other highly rated social values, for example, those of the market. Consistent with this, it remains that exercising control over things is taken to be the key to enhancing human well-being. It is a central organizing principle of modern society, a principal way to approach problems, accompanied by a confidence that further developments of our capabilities to exercise control will be able to deal effectively with any new problems and undesired side-effects that its exercise might create, as well as to introduce hitherto unthought of possibilities. Implementing a new technology may never be entirely free from controversy in the realm of values concerning, for example, whether its consequences will threaten civil liberties or produce environmental devastation. Nevertheless, modernity has unfolded with the confidence that risks are worth taking with technological advances, and that any problems that arise can and will be taken care of as the technological project itself progresses.

In the modern viewpoint technological advances as such, rather than, for example, the social relations under which technologies are implemented, are seen as the key to enhancing human well-being. That partly explains the acceptance of disruptions of social relations, and the imperative to construct new social arrangements, following the movements and implementations of technology. This gets at the sense in which exercising control over things has become considered as a value unsubordinated to other social values. The most distinctive modern value concerned with control, however, seems to be that of expanding the human capability to exercise control over nature.

The modern values of control

Values are held in integrated structured complexes (Chapter 2), the concrete manifestations of the items of which tend to reinforce one other. Modern value complexes include a set of distinct values about control, which I will call *the modern values of control*. Among the modern values of control, expanding human capabilities to control material objects ranks highest. Exercising control over material objects (as a characteristic activity of practical life so that, wherever possible, problems become redefined as having a technological solution) and especially implementing novel forms of control also rank high among these values. Thus technological objects and their products tend to be considered objects of value, at least some of them for some amount of time; and natural objects tend to be considered objects of value largely for their instrumental value. These values being held in a generally unsubordinated place, so that to their concrete manifestations the projects and institutions which express competing values must in large measure defer and adapt, may also be regarded as among the modern values of control. They gain reinforcement from the fact that they tend to be manifested in the same institutions as other social values (for example, private property, the market, and expanded options for choice) that are highly rated in modern value complexes; their manifestation also reinforces that of these other values.

The interests connected with the modern values of control can clearly be furthered by a form of understanding that enables us to encapsulate soundly what are the possibilities of control and the means to realize them in a way that abstracts them from their connection with lived experience, practical life, social arrangements and ecological and cosmic structures. This kind of encapsulation enables us to deliberate about control without being encumbered by considerations related to other social values, and to separate the question "Can it be done and how might it work?" from "What value does it have and is it worth implementing?" If these possibilities are to be implemented, however, they must ultimately be soundly represented *also* as functions of variables that can be directly manipulated by our action, under conditions which are within our power to bring about or to maintain, or which we have reason to accept will be present and maintained because of the way the world (or society) is.[1] The limits of possible control will be inseparable from the limits of this understanding. These limits, while they depend on human ingenuity in investigation and interaction with the world, cannot go beyond those set by the world. The world has been amenable to the expanded exercise of control that we have seen in modern times. But that leaves open whether the world may yet impose limits to the expansion; and under what social conditions such expansion may or may not occur. It also leaves open whether the expanded capability to control things (and its accompanying form of understanding) provides the key to enhancing human flourishing in all of its ideals under prevailing historical conditions.

THE MODERN VALUES OF CONTROL AND MATERIALIST STRATEGIES

Modern Western culture understands itself as the foremost bearer of rationality, and this self-understanding rests upon the twin pillars of science and technology. From one perspective, that in which *knowledge claims* are recognized as the primary location for rational evaluation, science looms larger. In its light the posits of theories developed under the materialist strategies, more than any other form of understanding, gain support from rational evaluative canons. They offer the best account we have of the nature and ways of material things, and consequently provide the theoretical underpinning of technological success and advance. From another perspective, that in which rational evaluation pertains in the first instance to *actions* designed to enhance the exercise of our designs upon the world, technology is in the foreground. Then science, conducted under the materialist strategies, gains rational precedence as a form of understanding because it provides the theory that furthers technological practice. Movement back and forth between the two perspectives is easy and common, since they are mutually reinforcing: the hegemony of technological practice is often grounded on its being informed by (materialist) scientific theory which is said to offer a superior understanding of the world; the virtual hegemony of research under the materialist strategies and the massive social and material investment in it are often legitimated in terms of its contribution to technological development. Either way the appeal to modern science carries an unmatched authority.

There is, in modernity, a mutually reinforcing relationship between gaining understanding under the materialist strategies and adopting the modern values of control,[2] marked by mutual dependence, many shared interests and conditions, intertwined causal dynamics, virtual inseparability of the social appeal and force of one from that of the other, considerable overlap in the institutions in which they are pursued – but less than complete identity. The materialist strategies and the modern values of control came into historical prominence together early in the seventeenth century, but they had distinct, earlier anticipations. Moreover, in the seventeenth century Bacon, expounding an early version of the modern values of control, advised (though he did not always practice) the use of inductive rather than materialist strategies; and Descartes, an early exponent of materialist strategies, did not justify their cognitive or rational merits (as distinct from their social importance) by appealing to their links with control. As ideas they are distinct and their historical dynamics have a measure of independence. Thus, for example, not every theory that is soundly accepted of a domain of phenomena under the materialist strategies can inform the expansion of our capability to control nature; and fundamental theories provide understanding of some phenomena that belong neither to technological nor to experimental spaces, those in which paradigmatically we exercise control over natural objects. Conversely, not every technological innovation reflects the input of (materialist) scientific understanding.

While the materialist strategies and the modern values of control are *distinct*, in the modern context it is difficult to *separate* them. As we will see, following the materialist strategies in research also contributes to the deeper manifestation of the modern values of control; and adopting these values with commitment motivates and depends upon furthering research under these strategies. The contributions are variegated and go both ways, not in every individual case, but as a solid, constant pattern, and not always in ways that are foreseen and antecedently intended. The strategies and the values have mutually reinforced each other so effectively that their interests largely have become identical under concrete historical conditions that make their simultaneous and reinforcing presence highly likely in leading social institutions.

The mutually reinforcing interaction between the materialist strategies and the modern values of control was anticipated by Bacon:

> I am laboring to lay the foundation ... of human utility and power. ... For the end which this science of mine proposes is the invention not of arguments but of arts; ... the effect ... being ... to command nature in action. ... nor can nature be commanded except by being obeyed. And so those twin objects, human knowledge and human power, do really meet in one ... [*The Great Instauration*]. ... now their understanding is emancipated ...; whence there cannot but follow an improvement of man's estate and an enlargement of his power over nature [*The New Organon*].
>
> (Bacon 1620/1960: 16–9, 29, 267)

It may be summed up in the following propositions:

1   The furtherance of the modern values of control is dependent on the expansion of understanding gained under the materialist strategies.
2   The pursuit of materialist understanding fosters an interest in the fuller manifestation of the modern values of control.
3   Materialist understanding is gained from the perspective of control.
4   Any values furthered by the pursuit of materialist understanding (for example, those associated with "fundamental" research) are manifested today as part of value complexes that also include the modern values of control.

In the next four subsections respectively I will elaborate these propositions; and then, in the following one, I will discuss links between the mutually reinforcing interaction and materialist metaphysics.

The modern values of control and the need for materialist understanding

A value complex cannot be manifested to any significant degree in practices and institutions unless the world is a certain way; in particular, the modern values of control cannot be manifested unless the world is amenable to being controlled by human action. What must things be like if they are to become objects of possible control for us? How should we aim to understand them if we wish to encapsulate extensively the possibilities of control of things open to us?

Suppose that we want to act upon an object, X, in order to bring about a state of affairs, S, where S is characterized abstracting from its place in human experience and practical life. To do this we will need to possess certain kinds of knowledge and certain skills and abilities. Concerning what we need to know, the following requirements (*requirements for control*) must be satisfied. We can identify conditions, $(C_1, C_2, \ldots, C_n)$, such that *given them*: 1) S's occurring can be represented as a function of X's gaining a property, P; 2) we can make X gain P by direct action; 3) each C_i is such that either a) its occurrence or maintenance is within direct human power, b) it can be controlled by a similar but independent process to that involved in bringing about S, or c) we have good reason to accept that it is a standing condition in the context; 4) S's occurring being a function of X's gaining P may reflect a lawful connection (or an empirical regularity) between S and X's being P, perhaps mediated by several lawful connections which may only be recognizable when underlying structure and process are considered, or that X's becoming P initiates a process that, barring further intervention, will eventuate in S.

S's occurring is within our capabilities of control if there are objects that we can manipulate in accordance with some set of requirements for control; and X is within these capabilities if there are states of affairs that we can bring about deliberately by manipulating X. The requirements include knowledge of *regularities:* given $C_1, C_2, \ldots C_n$, X's gaining P will eventuate in S; and identifications of what one can bring about by direct action, the immediate effects of one's bodily movements. The latter depend on personal know-how and ingenuity, cultural traditions, and the material and social organization of one's society. Only those established regularities, in connection with which we have the relevant control over the C_i and direct control over X becoming P, occur in sets of known requirements for control. Whether or not a regularity can inform our current capabilities to control material objects, however, it can always serve as a focus around which to explore extensions of these capabilities by indicating relevant conditions and things which, if brought under our direct control, would enable us to control further specified states of affairs. The totality of established regularities, thus, represents the limit of our capability to control things at a given time; so that a form of understanding that enables us systematically to derive and subsume regularities encapsulates (in the limit), by representing how to bring them to realization, the possibilities of things insofar as they are under our control.

Within the many forms of traditional knowledge, various sets of requirements for control have been established, so successfully that, for example, they informed the practices that produced the varieties of seeds, without which many current developments in agricultural technology would be impossible (Chapter 8). They have been derived inductively in the course of engaging in practices and skillfull activities in a familiar locale whose particularities, regularities and complexities of relationship have been charted through repeated observation and carried through the generations; but these sets have been limited in number, focus, systematicity and generality.

Understanding gained under the materialist strategies generates an expanding array of regularities with ongoing novelty of focus. A regularity may be obtained, directly in the course of experimental investigation, by induction from phenomena observed in experimental situations; from posing questions about how to extend our current powers to control; or more commonly by derivation from consolidated posits about underlying structure, process and law, posits which serve to explain, under the specified conditions, the connection between X becoming P and S. Often, therefore, objects as grasped under the materialist strategies are grasped in the way they need to be grasped to be included among those that lie within our capabilities of control, and they become objects of control most effectively in experimental and technological spaces, those spaces in which we initiate events under boundary conditions which close off other interferences so that the consequences of the initiated event all, as it were, flow out of the underlying order.

So if things are, or can become like, the way they are represented under the materialist strategies, they can become objects of control – provided that we can directly manipulate the relevant initiating events of regularities encompassing them, and ensure that the relevant boundary conditions remain in place, either through our own direct control or because they are standing conditions of nature as attested to in soundly accepted theories. The more general the laws are and the more wide-ranging the soundly accepted theories, the greater are the number and range of regularities, and (provided appropriate social arrangements prevail) of sets of requirements for control, which can be expected to be derived. I leave aside for now whether or not the sets of requirements for control established in traditional forms of knowledge can all be rearticulated within materialist understanding, and whether or not traditional forms can be developed through research so as to generate further such sets (Chapter 8). Whatever the case may be, materialist understanding leads us to sets of requirements for control that far transcend traditional constraints, so much so that in modern practical life, shaped as it is by the modern values of control, it has come about that objects, insofar as they are to become objects of control, tend to be considered as objects of materialist understanding.

Thus, holding the modern values of control brings with it an interest in the pursuit of understanding under the materialist strategies and of no other forms of understanding of material things. Under the materialist strategies we come to grasp material possibilities of things, which include (but go beyond) their possibilities as objects of control. Since understanding gained under the materialist strategies is wide-ranging (Chapter 5), and the ranges of which it provides understanding include phenomena that are within our direct control, grasping more of the material possibilities of things leads almost inevitably to grasping more of the possibilities of things as objects of control. It follows that the interests of the modern values of control are served generally by investigation conducted under the materialist strategies ("fundamental research"), and not just by that addressed to immediate practical problems ("applied research").

Materialist understanding (and, in the mainstream of the advanced industrial countries, largely it alone) grasps objects as they must be grasped in order to become objects of control. Since it abstracts from the relationships of objects to experience and practical life and to the social (and cosmic) order, it also encapsulates possibilities in a way particularly well suited for expanding our capabilities of control without subordinating the modern values of control to other social values. While it does inform practices that serve other social values (for example, of the market or the military), it does so only in virtue of particular instances of control becoming embedded in those practices. Under the materialist strategies phenomena are grasped only as generated from the underlying order or as subsumed under regularities, and so they are represented as of the same general character regardless of how they may be valued in the light of various competing value complexes. This is a mode of representation that is (by design) indifferent to differences of valuation, and so one in which moral reserve about certain practices of control cannot be expressed. It contributes to sustaining the sense, widespread in contemporary society, of the inevitability of someone (some corporation) bringing to realization whatever is shown to be possible (for example, connected with cloning and genetic engineering), and thus of technological "progress" into ever more realms of life and the ready tolerance of risks that generally accompanies it.

Reciprocity of theoretical and technological interests

The pursuit of materialist understanding fosters an interest in the fuller manifestation of the modern values of control. This interest derives in the first place from the facts that theoretical developments under the materialist strategies depend in crucial ways upon technological innovations which are only available where the modern values of control are deeply manifested, and second that where those values are deeply manifested it is virtually assured that theories established under the materialist strategies will be significant, needed to understand important objects in the realm of daily life and experience and to inform practical activities.

Developments under the materialist strategies depend upon the availability of technological innovations, increasingly of the most advanced and sophisticated kinds (themselves often products of applications of materialist understanding) to provide the necessary instruments and equipment to conduct relevant empirical (for example, experimental in the case of subatomic physics, observational in the case of outer space) and theoretical (for example, concerning computational needs) inquiries. Sometimes technological innovations are made in the first instance for the sake of furthering materialist scientific investigation. Then we may expect practical "spin-off" (for example, from computer software developments) simply from engaging in the research efforts as well as from subsequent practical applications (if there are any) of theories (for example, of high energy physics) that are consolidated in the course of the research – clear witness to the reciprocity and the dynamic interaction of the interests of materialist understanding and the modern values of control.

Technological advances today not only provide essential means for the progress of materialist understanding. They also may open up access to hitherto unknown, uncreated or inaccessible phenomena, or offer models (for example, the mechanical clock in early modern physics, the digital computer in contemporary cognitive science), without which certain phenomena would remain intractable to investigation, and so provide the occasion for defining new theoretical problems. In these situations, technological objects are an integral part of the research; they are among the objects of investigation.

Because of the dependence of its theoretical upon technological developments, materialist investigation can be conducted today at its most advanced levels only where the modern values of control are deeply manifested. Furthermore, as these values have become progressively woven into the predominant institutions of society, the more our practical activities and our daily life and experience in general become dominated by objects which are products of technology and thence explicable in their workings by materialist understanding. Materialist understanding has informed the social practices, expressive of the modern values of control, that have shaped the "world" of daily life and experience many of whose key objects can only be grasped and successfully dealt in its light. Thus, the significance of theories consolidated under the materialist strategies is virtually assured within these institutions.[3] Furthermore, for wont of their exercise in these institutions, our sensibilities become dulled to other forms of understanding, including those that might legitimate subordinating control to other social values, so that increasingly materialist understanding appears as the only form of understanding, as in principle without competitors, ensuring even more its significance.

It is when we attend to technological objects and to their place in the modern "world" (Chapter 7) of daily life and experience that the reciprocity of the interests fostered respectively by pursuit of understanding under the materialist strategies and by commitment to the modern values of control is most apparent. Simultaneously, in this "world" these objects are objects of materialist understanding whose existence in many cases is a causal consequence of practices informed by materialist understanding, and objects of high value from the perspective of value complexes that contain the modern values of control. It is little wonder, then, that in many research institutions and widely shared public conceptions the interests of the scientific and the technological tend to be considered effectively identical.

Materialist understanding: that grasped from the practices of control

Soundly accepted theories manifest the cognitive values to high degrees in relation to the appropriately selected set of empirical data. Typically, in research under the materialist strategies, these data are obtained from observing phenomena in the course of experimental practices, which are exemplary practices of control. We regularly anticipate, therefore, being able to generalize from them to further practices of control (for example, technological ones); and it is not uncommon for experiments to be performed for the sake of exploring technological possibilities and their effects, when the experimental space serves as a kind of a mini-prototype of a proposed technological innovation.

In saying that experimental phenomena occur within practices of control, I am describing them as products of human intentional agency, and as having humanly relevant consequences. Within the experimental practices, however, we describe the phenomena with categories drawn from the materialist lexicon, as we also describe the boundary conditions of the spaces in which they occur. Intentional agency stops short with fixing the boundary conditions (which may involve creating a complex and sophisticated space) and intervening to bring about the initial conditions; it then picks up again in order to observe and measure the experimental outcomes. This enables the phenomena of interest to be described adequately in materialistic terms and explained well in terms of the underlying order. Again, we anticipate being able to generalize from phenomena in such spaces, grasped materialistically, to similar phenomena and spaces including natural ones many of which are not and cannot be objects of human control.

Under the materialist strategies, making such generalizations is crucial for gaining adequate understanding of phenomena in natural spaces. While these phenomena may (and must) be grasped initially by observation and measurement, as well as a modicum of order arrived at through inductive inference or statistical analysis, in order to represent them in theories with significant explanatory power they must be represented with categories that have been shaped in the course of experimental (and measurement) practices and the theoretical efforts to make sense of the phenomena encountered in them. The language and posits of theory, even in cosmology, draw from experimental (and measurement) practices.

The heart of modern science is experimental. Experiment, as it were, sits between technology and natural spaces, providing for both a basis from which to generalize, a model of the ways things are, and a context for critical testing. Like technology, experiment is a human practice of control. Like phenomena in certain natural spaces, experimental phenomena can characteristically be portrayed as generated from underlying structure, process and law. In experiment, we come to identify or to confirm the powers of nature that we are able to deploy for the exercise of control over things.

Thus, even though materialist understanding extends well beyond the realm of control, and many scientists may value it largely for this reason, it is proper to identify it as understanding gained from the perspective of control. It is understanding of objects of the world insofar as they can be grasped from the perspective of practices of control. That understanding, it turns out, provides a sound grasp of the causal structure of numerous phenomena in spaces where human agency is not relevant, and it is broadly an empirical question how far such understanding can extend. The practices of control, in a many layered way, provide essential viewpoints, means and conditions for the pursuit of materialistic understanding.

"Fundamental" research and the modern values of control

There are often good reasons to engage in research in particular fields where there can be no reasonable general expectancy that established theories will become practically applicable. Clearly not all worthwhile research is motivated by the quest for practical applications, or concerned that any regularities it consolidates become items in sets of requirements for control. Historically, some phenomena have been the focus of spontaneous, culturally widespread and persistent interest because they pertain to reflections about "our place in the universe" or about recurrent and striking features of objects that impinge universally on human experiences. Any theories offering posits about the underlying order of these phenomena are normally assured of being highly valued, not because of their potential role in practical life, but because of the sheer interest of the domains. Consider cosmology. Here, one's interest is likely to be "knowledge for its own sake", knowledge of features of the world rather than knowledge intended for application in practices of control. Furthering the general objective of science, encapsulating the possibilities open to a domain of phenomena and discovering how they are realized is a value even if usually (Chapter 5) furthering it is bounded by the values that support adopting the strategies of a particular approach. In the case of cosmology, however, one adopts the materialist strategies for the sake of gaining understanding of phenomena in cosmological domains; adopting them is subordinate to the value of gaining understanding of these domains – and this approach has proved itself unrivalled in being able to generate theories that highly manifest the cognitive values of these domains.

In the previous chapter, I raised the question: Why are the materialist strategies adopted among the community of scientists virtually to the exclusion of any other strategies? Leaving aside the variety of individual motivations there may be for adopting these strategies, there could be a number of relatively distinct answers all pointing in the same direction. One answer, it will now be clear, rests upon commitment to the modern values of control: its interests are well served by research conducted under the materialist strategies as a whole, and not just by research directed immediately to informing particular controls.

There is also a second answer: the materialist strategies are adopted because they, and apparently they alone, enable the furthering of understanding, sought for its sheer intellectual interest, of certain domains of phenomena (for example, cosmological) or in "fundamental" research (for example, in particle physics, biochemistry, genetics, developmental biology or neurophysiology) that directly aims to grasp the underlying law, structure and process of phenomena in a deeper and more encompassing way? And does not this answer provide a more satisfactory explanation of the adoption of the materialist strategies since it fits better with the self-understanding of the scientific community, whose common articulations maintain that the cognitive and theoretical interests of science far transcend any links with the practical, and that any such links are effects of the successful and autonomous pursuit of these interests? Those attracted by the second answer might go on to suggest that the first is really an answer to a different question; not to: "Why has the scientific community adopted the materialist strategies virtually exclusively?", but to "Why do the relevant social institutions provide support for research conducted under the materialist strategies?"

The second answer cannot stand firmly on its own. It remains that any understanding gained under the materialist strategies, of any phenomena or of the underlying order, is gained from the perspective of control; and adopting the strategies requires certain material and social conditions which are products of implementations of the modern values of control. Within a value complex, understanding of the kinds of phenomena discussed on p. 125 may be held as an object of value; but any value complex which does include it as an object of value, assuming that it also accords value to the research practices from which the understanding is gained, includes also the modern values of control. That object of value, the understanding of these phenomena, can only come into existence where the modern values of control are highly manifested. Unless the value of the understanding gained is dissociated from the values expressed in the practices in which it is gained, and unless social values in general are subordinated to it, the interest of a domain of phenomena – considered apart from the mutually reinforcing interaction of the materialist strategies and the modern values of control – cannot provide a distinct ground for the almost exclusive adoption of the materialist strategies.

Nevertheless, the second answer does significantly supplement the first by pointing to how values, not reducible to the modern values of control, are also furthered by research under the materialist strategies. This helps to explain that, when we attend primarily to the immediacies of practical application or even to the general possibilities of control, the impulse behind much materialist research cannot be grasped – whether that impulse be connected with an intellectual interest in certain domains of phenomena or with the desire to create and consolidate "fundamental" (materialist) theories that manifest the cognitive values ever more highly as they unfold and when they replace each other. The grounds for the almost exclusive adoption of the materialist strategies leave open the focus of research activities under those strategies – whether it be "applied," focused on questions of immediate practical pertinence; or "fundamental," driven by the interests of obtaining theories that manifest the cognitive values ever more highly, and theories that are consolidated of particular favored domains. Either way it serves the interests of the modern values of control. My claim is that the mutually reinforcing interaction between the materialist strategies and the modern values of control explains their virtual inseparability in modern societies. But there remains a distinction between them, so much so that in "fundamental" research attention to the values has no role in the detailed play of the strategies. (One is apprenticed directly into the play of the strategies without necessarily gaining a clear awareness of the conditions that sustain the play, the values linked with their widespread adoption, and the way in which there might be alternatives to them.) Major scientific advances made under the materialist strategies (including those that are accompanied by radical switches in versions of the strategies, for example, from determinist to probabilistic versions) need not be occasioned (and generally are not) by any reference to control or to the values of control, but simply by the desire to gain theories (fitting the materialist strategies) that manifest the cognitive values more and more highly, or that manifest them of domains of phenomena of special interest.

Commitment to the modern values of control is the key to explaining the virtually exclusive adoption of the materialist strategies in modern science. Research under these strategies serves the interests that spring from these values – in general – and not only when it is addressed immediately to practical questions of control. In addition, any other values that are implicated in modern research activities or embodied in research institutions must, under modern historical conditions (whether or not articulated by the individual scientific investigator), co-occur in value complexes along with the modern values of control.

The relevance of materialist metaphysics

The mutually reinforcing interaction between adopting the materialist strategies and holding the modern values of control is further strengthened when the adoption is considered to be grounded in the acceptance of materialist metaphysics. Materialist metaphysics affirms that the world "really is" such that all the objects in it (including human beings) are fully characterizable by materialist (perhaps ultimately physicalist) properties and relations and all phenomena in terms of being generated in accordance with underlying structure, process and law; and the possibilities of things are exhausted by their material possibilities. Then, in principle, following the materialist strategies could give us a complete account of the world. In principle, no possibilities would be left out. I argued on p. 120 that gaining access to the material possibilities of things contributes to expanding our capability to exercise control over material objects and states of affairs. If, essentially, there are no other possibilities then gaining understanding of the world *per se* contributes to expand this capability. In addition, regardless of what projects of control we may incorporate material objects into, their nature remains unchanged; not being objects of value *per se*, it is open to us to accord them whatever value we desire and to deny them any but instrumental value – control cannot be opposed with the argument that it changes the natural character of things.

The very nature of the world, then, seems to underlie the exercise of control as the characteristic activity to engage in when relating to material things; and *ceteris paribus* gaining understanding of it expands the range of possibilities for its exercise. Furthermore, not only do the categories deployed under the materialist strategies not provide a barrier to moving from the possibility to the legitimacy of introducing particular controls, but (in principle) they are the ones appropriate to grasp the world as it is, implying that categories (value ones) that one might use in attempts to bar the move have no grip on the world. It also supports that, where practices of control bring with them undesirable or unexpected side-effects, in principle they can be dealt with through further controlling interventions.

Although this story has a certain appeal there are tensions in it. Like all stories it is told in intentional not in materialist idiom. Our understanding of the gaining, consolidating and applying of materialist (including physicalist) understanding is expressed in intentional categories, as is our understanding of human action in general. This mode of understanding is linked with the value that human beings not be treated as objects of control. Agents are not, by nature, objects of control, though (by diminishing their agency) they can be made to approximate them (Lacey and Schwartz 1986; 1987). The exercise of control (by humans) over humans diminishes their agency, an essential aspect of human nature (cf. Chapter 9), and this underlies a general moral objection to establishing relations of control among human beings (Lacey 1979; 1985; 1990). But agents are part of nature and so it is appropriate to expect that a general view of nature be formulated with categories apt for representing agency – especially since agency is both a phenomenon of lived experience and practical life, *and* a presupposition of scientific practice.

The content of materialist metaphysics derives from extrapolating the categories of theories, well established under the materialist strategies, to all phenomena and states of affairs in the world. There is no compelling reason, however, to hold that the very activity that produced the theories in the first place can be adequately represented within the theories' own categories (no matter how they may be generalized and abstracted). Perhaps it can be; but it has not (yet) been. Meanwhile, I see no serious difficulty in treating intentional understanding effectively as neither reducible nor replaceable, though often needing to be supplemented. I said that accepting materialist metaphysics strengthens the mutually reinforcing interaction between adopting the materialist strategies and holding the modern values of control. Early in the modern scientific tradition attempts were made to ground versions of materialist metaphysics a priori. Most agree that these attempts failed though their residue remains, and many suggest instead that it is grounded dialectically as both an extrapolation from and a presupposition of the success of modern science.

But materialist metaphysics is not a presupposition of the remarkable success of modern science in generating theories that come to be soundly accepted in the course of following the materialist strategies. It is enough that there exists a wide (and, in principle, unlimited) array of spaces – many of them created by human experimental and technological intervention – in which phenomena can be well represented as generated from the underlying order. Parts or aspects of the world must be this way if we are to come to accept the theories we do accept. And so, too, must we be intentional agents. Nothing more need be presupposed; certainly not that human action can be understood in terms of the same kinds of principles deployed to understand phenomena in these spaces.

What can be extrapolated from the success of science? Certainly that more phenomena in more spaces will fall under the grasp of materialist understanding, that the laws represented in widely encompassing theories may represent universal tendencies of nature (though not that universally they are highly salient explanatory factors), and that many of the entities discovered as a consequence of experimental activity have important effects both in the natural world and in the "world" of daily life and experience. We can also extrapolate that increasingly material possibilities of things will be encapsulated, so that increasingly things will *become* objects of possible control. But there is no sound inference from *can become* to *already is*, or to *cannot become otherwise*. The content of materialist metaphysics can be extrapolated from the most soundly accepted, widely encompassing theories, but an argument to endorse this metaphysics cannot be.

What, then, explains the appeal of materialist metaphysics, the commitment or certitude – going well beyond what can currently be established by evidence or argument – displayed by those who adopt it, and their confidence that difficulties will be overcome and that particular refuted arguments of theirs will readily be replaced by better ones? Perhaps it is the allure of a unitary world view. Perhaps it is a sense that unless the world can be grasped under the materialist strategies, we cannot have theories manifesting the sort of "clearness and distinctness" needed to bring them into a decisive meeting with the empirical data, and thus to have genuine knowledge. Relatedly, perhaps it is a sense of the ultimate unintelligibility of alternatives; and that despite the difficulties that intentionality imposes for materialism, there remains an on-going program that seems to be making progress in dealing with them. Perhaps it derives from the mutually reinforcing interaction between the materialist strategies and the modern values of control, that the way of understanding reflected in the interaction has so come to dominate our consciousness in practical life that no other mode of understanding seems comparable in power, to be intelligible, or even to be worthy of exploration.

Whatever the explanation may be, it falls short of providing a compelling argument for the adoption of materialist metaphysics. I incline to the last proposal. Then, materialist metaphysics does not provide an argument for the adoption of the materialist strategies that is independent of the one rooted in its mutually reinforcing interaction with the modern values of control.[4]

This conclusion accords with my general view of the relationship of metaphysics with science. Like the empiricists, I do not think that science, defined by the general objective to gain understanding of things (O, Chapter 5), is committed, except heuristically and temporarily (Hesse 1977), to any particular metaphysical view – either as presupposition of its practices or as implication of its established results. Like the scientific realists, I think that there are posits of the underlying structure, process and law of certain spaces that are so well confirmed as to be placed in the stock of uncontested knowledge (cf. McMullin 1998: 378). An approach to science proceeds under particular strategies, whose source need not be in metaphysics but can be, as I am arguing, in values. For instance, it need not be in considerations about the general nature of things, but may be in considerations about the general possibilities of interest for our (stances of) interacting with the world.

Looked at this way, we can recognize the most general categories of the lexicon borne by the materialist strategies to be derived not from materialist metaphysics, but from responding to the question: "How must we think of material objects if we want to further the manifestation of the modern values of control?" They appear to be derived from materialist metaphysics – and the lexicon appears to be the lexicon of *science* rather than of a particular *approach to science* (Chapter 5) – when commitment to it is unproblematized and the modern values of control are uncontested. Then the ceaseless expansion of our capabilities of control also appears to be ensured by the metaphysics. When we attend to the lack of solid grounding for materialist metaphysics, however, my conclusion that its grip comes from the mutually reinforcing interaction of the materialist strategies and the modern values of control becomes compelling.

CONTROL AND UNDERSTANDING OF "THE MATERIAL WORLD"

My answer to the question: "Why adopt the materialist strategies to the virtual exclusion of other strategies?" is neo-Baconian. Control of nature is the key. Not that the aim of science is the control of nature or that (fundamental) research projects are shaped by immediate practical concerns; but that the modern values of control have become deeply woven into modern society and its most powerful institutions, and that they interact in mutually reinforcing ways with research conducted under the materialist strategies, whose theoretical products are generally significant in the same institutions as those into which the modern values of control are woven. Consistent with this, it remains a necessary condition for the adoption of these (or any other) strategies that under them are produced theories that manifest the cognitive values highly. But, the interest to produce theories that manifest the cognitive values highly cannot explain the virtual exclusivity of adoption of the materialist strategies. Cognitive interests underdetermine the choice of which strategies to adopt, and social values take up the slack. I emphasize that cognitive and social values do not play their roles at the same level. The social values provide an important part of the reason to adopt a strategy, but theories developed under the strategy are properly accepted in virtue of the manifestation of the cognitive values. My account preserves *impartiality* as an ideal of scientific practice, though it cannot remove *neutrality* from ambiguity (Chapter 10).

My answer to the question in the previous paragraph clashes with the idea that science is value free (with *autonomy* and perhaps *neutrality*) so that those who articulate the modern scientific tradition will be reluctant to endorse it. Their reluctance will be assuaged neither by my preservation of *impartiality*, nor by my claim that providing special support for values, widely assumed to be universal and partly constitutive of rationality, is the key ground on which *neutrality* appears to be violated. The reluctance derives from the conviction that modern science, the project that has vastly expanded our common stock of knowledge, is driven (ideally) by purely cognitive interests and that these suffice to explain not only the proper acceptance of theories, but also the sound adoption of strategies.

The reluctance is compatible with recognizing that there is the mutually reinforcing interaction between adopting "the strategies of modern science" and the modern values of control, that this interaction may provide the motivation for individual scientists to engage in research, and that it serves to explain the ready availability of the material and social conditions required for research in the advanced industrial countries. Then it might be objected that all I have succeeded in explaining is the accelerating pace of scientific development and the widespread social support for the activities conducted by the scientific community, not the grounds for the almost unanimous adoption of the materialist strategies within the scientific community. Perhaps, the objection might go, I have rushed too hastily to conclude that cognitive interests underdetermine the rational choice of strategies. In particular, that I have not sufficiently considered that our capabilities to control material things have been successfully enhanced (and thus the mutually reinforcing interaction between the strategies and the modern values of control has become possible) because research conducted under the materialist strategies enables the consolidation of theories, more and more of them encompassing ever larger domains of phenomena, with greater cognitive credentials than understanding gained from other strategies. If this is so, then could we not explain the almost unanimous adoption of the materialist strategies in terms of them being the strategies under which we can consolidate theories that manifest the cognitive values most highly of whatever domain of phenomena is chosen for investigation – and would not that explanation stand on its own regardless of whatever reinforcing social explanations there might also be?

Is this so? It is true that theories consolidated of certain domains under the materialist strategies are commonly considered to be exemplars of items properly included in the stock of knowledge. But that is not enough to sustain the objection that the cognitive values do not underdetermine the choice of strategies. It would have to be argued that in *principle* theories consolidated under the materialist strategies manifest the cognitive values most highly, and not just *in actual fact*, for the prevailing social conditions might account for the underdevelopment of alternatives (Chapters 7 and 8). How might such an argument unfold? There is a hint in the suggestion just made that our capabilities to control material things have been vastly enhanced *because* research conducted under the materialist strategies has consolidated theories with superior cognitive credentials. Control, we might say, has been enhanced because we have gained more and better knowledge of material things; enhanced control is a symptom of the cognitive superiority of theories consolidated under the materialist strategies. An argument of this kind has been proposed by Taylor (1982).[5] I will now offer a version of it.[6] In it "comprehensiveness" is the key cognitive value appealed to; understanding gained under the materialist strategies is maintained to be inherently the bearer of the most comprehensive understanding of the "material world" or the "physical universe."

The value of materialist understanding

To understand a thing, according to Taylor, is to have a "rational grasp" of it, to have an "articulation" of it in which its various features are distinguished and presented in "perspicuous order," so that the cognitive or rational credentials of an articulation derive (provided that it is consistent) from the perspicuous order that it lays out (90). Taylor is sensitive to the intelligibility of alternative forms of understanding (Chapter 5) conforming to a variety of ideals of "perspicuity." Nevertheless, he proposes that greater perspicuity derives from "a broader, more comprehensive grasp on things" (ibid.), and it is such a grasp that is provided by understanding gained under the materialist strategies.

How is it that materialist understanding provides a broader, more comprehensive grasp of things? I distinguish this question from: Why is it that materialist understanding has displaced (for the most part) hitherto existing forms of understanding (pertaining to material objects)? My account of the grounds for adopting the materialist strategies answers the second question. For Taylor, in contrast, it is the comprehensiveness of materialist understanding that principally accounts (rationally) for its having displaced earlier alternatives, though comprehensiveness functions in concert with an account of how materialist understanding and earlier forms of understanding are incompatible, how they cannot both generate generally significant products in the same social/historical/cultural nexus.

In an earlier phase of Western civilization there was a form of understanding in which "understanding" and "attunement" were inseparable, and in other cultures there are forms of understanding in which the modern Western separation between practical activity and symbolic expression cannot be made (Taylor 1981: 209). Taylor maintains that they are incompatible with materialist understanding. It is not that their products are formally inconsistent (Chapter 7); rather the incompatibility derives from the ways in which they inform human action and the practices from which they are produced. We may say that to each form of understanding there corresponds a characteristic activity, or predominant stance towards nature, which not only does the form of understanding illuminate, but which also contributes towards the generation or consolidation of the form of understanding. For example, where understanding and attunement are inseparable there is the characteristic activity of adaptation to nature, and corresponding to materialist understanding there is that of exercising control over nature framed by the modern values of control. While, in any culture, elements of the various characteristic activities may be present (at least marginally), as predominant stances those of control and, for example, adaptation mutually exclude each other. They do not fit together or complement each other; they "are rivals; their constitutive rules prescribe in contradiction to each other. ... [They] cut across [each other] in disconcerting ways" (98–9). They are not merely different; they cannot be adopted together, and the attempt of different groups to adopt them simultaneously in the same social space must lead to conflict. The very success in exercising control changes the environment in which one lives, yet adaptation presupposes a more or less constant environment, subject at most to periodic rhythms. Conversely, the predominance of adaptation precludes the kind of exercise of control necessary for the conduct of research under the materialist strategies (Chapter 7).

Where the stance of adaptation has been adopted as the predominant one of a culture its associated form of understanding is taken both to have empirical support and to articulate the value of the prevailing social (ecological, cosmic, and possibly spiritual and theocentric) order; it makes sense of a good deal of daily life and experience, and outlines the path to attunement. It also delimits the class of the possible: it may lack the conceptual resources to encapsulate the possibilities that may derive from another stance, and its posits will not have been tested against those that might be put forward by a form of understanding associated with a different stance; and it denies the possibility of attunement if one engages in a different activity. There is a sort of self-enclosure that neither recognizes nor permits much space for adopting alternative stances – except on the margins – and possibilities that are not identifiable and realizable within it, even if recognized, are not valued. Similarly, the modern Western "enclosure," defined by the shifting perspectives of (materialist) scientific understanding and technological control, and thus not dependent on stability, but on constant innovation and change, leaves little space for alternative activities and for the exploration and identification of possibilities that are not realizable within its structures (Chapter 8; Lacey 1997c).

In order to challenge the demarcation of the range of possibilities that a form of understanding associated with adaptation admits, one would have to engage in activity under an alternative stance – but that would threaten attunement. Therefore, one cannot rationally challenge the demarcation of possibilities without also challenging the valuation of them. One must engage in that activity (which is negatively valued within the prevailing form of understanding) *prior to* gaining empirical evidence that there are genuine (and valuable) possibilities that are unrecognized within the self-enclosure. In doing so, one might threaten the viability of the stance of adaptation for the culture as a whole, since the natural world in which adaptation takes place could be significantly changed in the course of attempting to establish one's claims. For the alternative activity to get under way, it seems, there would have to be doubts about attunement. At least on the margins there would have to be embryonic grounds to suggest that the alternative practice may produce betterment for human beings. Otherwise, the self-enclosure would seem to be unbreakable, unless it succumbed to a natural disaster or to a powerful outside intrusion.[7]

The promise of human betterment was a major theme for virtually all the important contributors to the scientific revolution of the seventeenth century. It has been ignored in many discussions of the superiority of modern to ancient science, being held to be of no cognitive significance. Certainly it has no cognitive bearing on whether a certain theory best articulates a given domain of phenomena. But it does bear upon whether the pursuit of materialist understanding is desirable and whether the understanding gained is significant. In the process of the scientific revolution, the notion of "human betterment" like that of "theoretical understanding" was transformed. Taylor says that, in older traditions, not to be attuned to nature is "to be in misery and confusion." Through this negative side there is retained a thread of contact between the old notion of attunement and the modern sense of betterment.

Materialist understanding and the displacement of earlier forms of understanding

Human activity in all cultures involves some measure of exercising control over natural things. In a broad sense, all living things intervene in nature through various mechanisms of assimilation and accommodation. The human distinctiveness is that the intervention is purposive and planned, expressive of stances which are illuminated by forms of understanding. The difference between the stances of adaptation and control is that in the former, but not the latter, the exercise of control is subordinated to such values as ecological and social stability and bounded by ends and means defined in traditional practice (Chapter 8) or, in Taylor's terms, the activities in which control is exercised are not articulated as disjoined from "expressive activities" (Taylor 1981: 209). In contrast, when the stance of control is adopted the exercise of control is framed by the modern values of control; ends are not circumscribed as a matter of course by a natural environment which is to be nurtured or a social order to be maintained, but only by the possible and the power to implement it.

The differences are important. Nevertheless the forms of understanding of older traditions do have components that inform certain practices of control of material objects (in agriculture, engineering, medicine, etc.); they have a certain amount of systematic empirical knowledge of various sets of requirements for control; or they have understanding which informs our "ability to make our way around in [the world] and deal with things in it" (101). This enables the transcultural recognition of effective practices and their comparison and, thus, the transcultural recognition of the much greater practical efficacy of technological practices in which the stance of control is adopted. The scope, power and effects – penetrating, as they unfold, into the lives of almost all human beings – of these practices expand the horizons of the possible (and the historically actualized) in a way that far surpasses anything entertained in traditional forms of knowledge (Taylor 1995). Because of their continuity with the goals of many traditional practices of control, because of their greater effectiveness and efficiency in serving some of them and thus (at least piecemeal) their being welcomed as replacements for some traditional practices and sources for the betterment of the human condition, and because of their looming presence as influences confronted in daily life and experience, the achievements of technological control cannot be ignored. The mechanisms that account for this impact of technology, that cannot be ignored, can be complex and various and those that were important in the West have not generally been recapitulated elsewhere. Sometimes desire for the "betterments" promised by modern technology is a key factor, and sometimes the interests of powerful elites; at other times colonial force plays a role, but Taylor is right to emphasize the continuity of technological with traditional practices so that members of a culture can embrace a technological practice because it performs a traditional task more efficaciously.

Any form of understanding, which cannot encompass the possibilities and explain the material workings of technological controls, becomes unable to inform daily life and experience adequately, and so its theories (posits) are rendered insignificant. A form of understanding which cannot do this will, as the impact of modern technology grows in the daily life and experience of a culture, be displaced by one that can. Displacement occurs, I suggest, when a culture's contact with technology has been sufficiently prolonged that technological activity has become either highly salient or pace-setting in the practical life of the society. It does not seem to be possible in the contemporary world for a culture to isolate itself significantly from the encroachments of technological advance, so much so that technological activity has become a major factor in shaping the future of every culture, even if only to the extent of threatening to destroy the space within which a particular culture thrives.[8]

It is materialist understanding that encompasses the possibilities and explains the material workings of modern technological objects. This leads to an "inner connection" between understanding of "the world" and control which "rightly commands everyone's attention" (101, 103). Thus, materialist understanding displaces earlier forms of understanding, unless they are able to develop so as to enter into rich dialectical relations with it, as forms of understanding associated with the Western pre-seventeenth century stance of adaptation did not.

The displacement argument relates to considerations of comprehensiveness ambiguously. Its conclusion is not that materialist understanding can in principle encompass all phenomena of which other forms of understanding offer accounts. Materialist understanding does not explain, for example, the social forces and the sources of the social values that have come to be woven into social forms along with the modern values of control that must also be grasped in order to fully understand the achievements of technological controls. Rather, it encompasses key phenomena and possibilities, those of control characterized in abstraction from their social and cultural context, with a comprehensiveness unrivalled (and unapproachable) by its pre-modern competitors. It is a kind of "bounded" comprehensiveness that is involved – bounded by the interest of control. Without historical success in transforming the world of daily life and experience which derived from adopting the modern values of control, the displacement argument would have gained no footing.

The displacement argument, therefore, does not challenge my explanation for the virtually exclusive adoption of the materialist strategies in the modern scientific community.[9] On the contrary, it deepens the reflection on the reciprocity of interests between research conducted under the materialist strategy and adoption of the modern values of control. The phenomena and possibilities of control simultaneously are objects grasped within theories consolidated under the materialist strategies and objects highly valued within complexes that contain the modern values of control. The more salient their place becomes in a society the more powerful is the displacement argument, for a form of understanding must offer an account of salient objects in the world of daily life and experience; that is the test of the significance of the theories it produces.

This leaves open, however, that there may be good reasons – linked with subordinating control to other social values – to explore sources of alternative strategies in which the understanding of material objects does not involve abstracting them from their relations with human and social factors, and which thus explore possibilities that are not encapsulated under the materialist strategies. Such alternative strategies may meet the just-mentioned test of significance by absorbing materialist understanding into a subordinate place within a form of full understanding (Chapter 5) in which its limits of applicability (including desired applicability) would be identified, and in which more generally an account of materialist understanding and its place would be constructed, accounting for its trajectory within currently dominant social forms and characterizing what its trajectory would be in the alternatives to it. No doubt these strategies would compete with the materialist strategies for social space in which to develop, but not (normally) with them as generators of theories that encapsulate the material possibilities of things. Understanding gained under them would not challenge that theories that encapsulate the material possibilities of things regularly and progressively are consolidated under the materialist strategies, and that they represent a remarkable contribution to knowledge. It queries, not the knowledge, but its significance, and thence the value of gaining (without subordination to specified values) further materialist understanding. These conjectured alternatives may be continuous with older forms of understanding: they may result from the development and radical redeployment of older forms in the context of modern daily life and experience, involving dialectical syntheses of old strengths and new gains. There might even be important residue to rescue from the older forms concerning particular material objects and practices, something that would not be surprising given that many older practices presumably survived because they were soundly based empirically (see references in Chapter 8). The displacement need be neither a wholesale obliteration nor a barrier to the emergence of alternatives.

Technological objects and their gaining an important role at the "cutting edge" of Western society significantly pre-dated the capability of materialist understanding to anticipate novel material possibilities and to uncover the means to realizing them. Nevertheless, for well over a century now materialist understanding has contributed enormously to furthering the technological project and thus to bringing about a world, key objects of which have made recourse to materialist understanding indispensable and apparently ever more so, as technologies spread into more and more domains of life. That it has become indispensable in this way helps to explain why there are plentiful resources available to pursue research under the materialist strategies, but it does not provide a general reason (as distinct from one grounded in endorsement of the modern values of control) to adopt them in research rather than any others – except by default. The mark of historicity remains. The socio-historical world had to change for the materialist displacement to occur and to be rationally grounded. It is the success of a socio-historical project (itself informed by materialist understanding), the ever deeper manifestation of the modern values of control, that has made the materialist strategies indispensable for grasping "the world we live in" today.

In a similar way, an alternative form of understanding may come to inform a socio-historical project, perhaps articulated around the ideals of ecological soundness and social justice, that may become realized; so there is no less reason to explore such alternatives now than there was to explore materialist understanding at the very beginning of modern science. There are powerful social forces posed against the success of such nascent alternatives (regardless of whether there are good reasons to explore them), so that power – mediated through its being exercised on behalf of the interests of the modern values of control – may be an important part of the explanation for the dominance of the materialist strategies. That is consistent with my viewpoint, but not with those who maintain the unqualified cognitive or rational superiority of the products of research under the materialist strategies.

The comprehensiveness of materialist understanding

The fact that it is materialist understanding that encompasses the possibilities and explains the material workings and effects of modern technology is sufficient to account for the kind of displacement just described. Attempting to move beyond displacement, Taylor goes on to argue that materialist understanding is able to displace earlier forms of understanding because it offers a broader, more comprehensive grasp on things; "it has greatly advanced our knowledge of the material world" (103). Modern technological success, he maintains, depending as it does on our having gained understanding that encompasses the possibilities and explains the material workings and effects of technological objects, reflects the greater comprehensiveness of this understanding; the success of technology is "proof" of this. That it enables higher manifestations of a cognitive value, "comprehensiveness" – not that it successfully informs technological practices – it is thus proposed, grounds the cognitive or rational superiority of materialist understanding; technological success is a symptom not a ground of this cognitive superiority. Then, one might argue, it is the quest for this kind of cognitive superiority that explains the almost unanimous adoption of the materialist strategies within the scientific community. We might put it: adopt the materialist strategies because research under them can provide a more comprehensive grasp of *the material world* rather than because it can provide a better account of a range of possibilities (the material possibilities of things) that are considered especially valuable.

Does materialist understanding offer a more comprehensive grasp of things? Taylor says: "modern science represents a superior understanding of the universe, or if you like the physical universe. ... [It] achieves greater understanding at least of physical nature" (102–3). ... It is infinitely superior for understanding the natural world. Our immense technological success is proof of this" (Taylor 1981: 209). Does materialist understanding offer a more comprehensive grasp of the material (physical, natural) world? (What is the "material world"?) Does the fact that it explains technological successes and anticipates novel technologies – more generally, that it enables the encapsulation of the material possibilities of things in an ever more encompassing way – reflect that it offers a more comprehensive grasp of the "material world?" Is there an argument, independent of technological success, for this greater comprehensiveness? Or is the argument simply that a condition of the possibility of technological success is that the understanding which informs it must manifest such comprehensiveness?

Materialist understanding is wide-ranging; but it is not full (Chapter 5), since it abstracts from the human, social and ecological dimensions and consequences of phenomena within the spaces of which it provides understanding, and from the human and social possibilities that may be hidden in them. It encapsulates the material possibilities of things, those possibilities that can be identified from the generative power of underlying structure, process and law, abstracting from the place of things in human experience and practice. Only in spaces where human factors are causally irrelevant could it be seriously entertained[10] that the material possibilities be considered identical with the complete range of possibilities permitted by the space and the arrangements of phenomena in it; and only in those spaces where ecological and organism/environment interaction factors are irrelevant can one plausibly be confident of the identity.

Modern technological successes and anticipations of novel technological innovations presuppose only that the understanding that informs the workings of technological objects is wide-ranging and that the spaces over which it ranges include experimental ones. Technological objects, like experimental phenomena, fall within the domains of phenomena of which theories, consolidated under the materialist strategy, manifest to a high degree an array of cognitive values, including wide-ranging explanatory power. Modern technological success – not technological advance *per se* since it predated sophisticated developments of materialist understanding – attests to success in having gained wide-ranging understanding, and to the historical superiority of materialist understanding as an instance of wide-ranging understanding. Equally, however, the success of research under the materialist strategies presupposes that we have successfully exercised control in numerous experimental spaces and made certain kinds of technological advances. Materialist understanding and our capabilities to exercise control develop together in mutually reinforcing interaction in an unfolding spiral of development, so that each moment of development of the one presupposes relevant moments of development of the other.

Materialist understanding encapsulates the material possibilities of things, and helps to generate sets of requirements for control, in a more encompassing way than rivals. But it does not generate full understanding: it not only does not treat things as cultural objects (of course some things are not cultural objects) or their effects on human lives or the human causal factors which influence the boundary and initial conditions of their motions and changes and which may be essential for their very existence, but also does not deal centrally and sometimes not at all with the side-effects of technological interventions, including their social and environmental consequences. Materialist understanding deals with things solely under descriptions that relate them to underlying law, structure and process; a mode of understanding which underlies the power, given certain boundary conditions (which it characterizes without reference to the human agency and social conditions that may be necessary to bring them into being) to predict and control, and thence the power to uncover novel (material) possibilities of nature. Nevertheless, its mode of dealing with most of the objects we encounter in daily life involves abstraction. Theories consolidated under the materialist strategies manifest well the cognitive value "wide-ranging explanatory power," but not "full explanatory power." The gain is more or less unimpeded technological advance. One problem is that technological objects are commonly introduced without a full understanding, and this is reflected in current social and ecological crises. Another problem is that the mutually supporting interaction of research under the materialist strategies and the modern values of control remains implicit, without the sharp articulation that it might gain within a form of full understanding, thus disguising that much of materialist understanding is not *neutral*, that its general significance is assured only where the modern values of control are deeply manifested.

Perhaps I have side-stepped from the central thrust of Taylor's argument. In it such phrases as "the material world," "the physical universe" and "the natural world" frequently recur apparently as synonyms. It is of the "material world" that materialist understanding is said to give a more comprehensive account. Materialist understanding comes to the fore when we dissociate activities of control from expressive and valuative activities. That dissociation is historical, but when we make it we gain superior access to the "material" (or "natural") world which apparently is not considered as constituted within human history. Since Taylor subscribes to neither materialist nor Cartesian metaphysics, I find it difficult to interpret these phrases. Human beings are part of nature or "the natural world." But then there are parts of this world, those where human phenomena are pertinent, which are not well grasped under the materialist strategies; the comprehensiveness Taylor speaks of does not embrace these phenomena.

"The material world" and its synonyms are intended to designate the object of inquiry in the natural sciences, but it seems that it cannot be identified with "the natural world" that we are part of in all its fullness. This object of inquiry (Taylor says) "exists independently of us human percipients" or consists of bits of the natural world and how they affect each other "even when we aren't on the scene or we aren't playing a role" (32, 47). In this "world" human causal agency is not relevantly a factor, and its components and their properties (unlike those investigated in the interpretive human sciences) are not partly constituted by the interpretations of human beings engaged in various practices. Can we properly characterize the object of research conducted under the materialist strategies in this way? Any answer to this question, I suggest, must be consistent with the following conditions:

1   Our understanding of objects is mediated by available conceptual resources and (in scientific practice) by the lexicons of the adopted strategies.
2   We are part of the natural world, causal agents in it; we cannot grasp the world except in virtue of our causal agency in it.
3   We are able to grasp segments of the natural world, in which human agency is not a relevant causal factor, through practices that relate them to segments (usually experiments) where human agency is relevant.
4   We are able to extrapolate this grasp to spaces even further removed from human experience and agency, including to times before there were any human beings and, in principle, the extrapolation may extend to the coming into being of human beings with their distinctive capabilities, with the extent of the extrapolation remaining an open question, often begged by premature commitment to materialist metaphysics.

Taylor holds that the lexicon deployed within the materialist strategies reflects that the object grasped in inquiry under them is properly characterized as independent (ontologically) of human observers. He distinguishes "subject-related" or "anthropocentric" (Taylor 1985: 2; 1995) and "absolute" terms. *Subject-related terms* designate subject-related properties, those which a thing has in virtue of being an object of experience of human subjects, its situation within the context of human action, interest or concern, and how it affects human beings. *Absolute terms* are non subject-related terms; they designate properties that are not subject-related, properties that things have regardless of their relations with human beings. According to Taylor, the lexicon deployed within the materialist strategies consists entirely of absolute terms. So, we might put it, the "material world" is the totality of objects designated by the lexicons used to articulate the theories consolidated under the materialist strategies. This move shifts our focus to the terms contained in these lexicons. When we make it, however, I am led to a conclusion that diverges sharply from Taylor's. Taylor offers as examples of subject-related properties desirability characterizations of people, and the secondary qualities, those qualities (according to him) that things have in virtue of their powers to cause certain types of experience in people. The terms used in the materialist lexicons indeed have different features from these examples and, as deployed in theories, they show no manifest "sign" that they are subject-related too, and they are by design dissociated from value terms. Are they absolute?

Two (interrelated) considerations need to be brought to bear. First, these terms are predicated of things in the course of scientific (theoretical, measurement and experimental) practices conducted under the materialist strategies; and which terms are predicated (and which of alternative largely incommensurable lexicons they are part of) varies with changes within the materialist strategies. Many of them are predicated successfully; for instance, they enable the articulation of theories that highly manifest the cognitive values of a wide range of domains. This much is clear. There is nothing about the scientific practices, however, that enables us to conclude that the terms designate independently existing properties of things. The practices are compatible with there being no more to affirming that things have certain properties than that the relevant terms have been successfully predicated. Then the terms of the materialist lexicon would not be absolute. If so, and if the "material world" is that which is represented in our best scientific theories, then the things of the "material world" and their properties are partly constituted by the practices (and their associated lexicons) which enable their articulation.[11]

Second, the terms of the materialist lexicon are for the most part quantitative. Quantitative terms might be taken to be paradigmatically absolute; a quantity apparently reflects no values, and its applicability follows from procedures whose outcomes are intersubjective and shared among a wide variety of human practices that express many different interests. Things are not so simple. That a procedure leads to designating the same numerical value of a quantity to a thing regardless of who conducts the procedure does not guarantee that essential links with values are not implicit. The use of quantities in the social sciences makes this clear. (Think of the controversies about IQ and the index of unemployment.) A quantitative term is applied to particular things *via* some measuring operations and usually an array of mathematical inferences and calculations that are theory-dependent. Measurement requires instruments and bringing the thing to be measured into appropriate relationship with the instrument, so that it concerns a relation between the thing and the instrument, which is a human artifact, theoretically articulated, that we insert into the space of investigation (Chapter 7). The meaning of a quantitative term cannot be dissociated from the operations of instruments; nor can it be reduced to them, since it is also implicated in the theoretical articulations of the construction and uses of the instruments. Quantitative terms are not absolute; they do not refer to properties that things have regardless of their relations with human beings and of their role in human practices.

In the light of these two considerations I do not think that a relevant, ontologically grounded distinction between absolute and subject-related terms can be sustained. Where does this leave the "material world?" I think that appeal to it is largely the residue of Cartesian metaphysics. Leaving aside Cartesian or materialist metaphysics, one might consider it to be whatever is represented in the best theories consolidated under the materialist strategies, but that does not serve Taylor's purposes. Even so, it is a useful way to think of it. The "material world" is the world as grasped in terms of what can be generated from underlying law, process and structure by means, not of absolute terms, but of categories suitable for articulating underlying law, process and structure. This is effectively equivalent to the world as grasped from the perspective of the stance of control. Alternatively, and again equivalently, we may think of the "material world" as the totality of the material possibilities of things. Theories consolidated under the materialist strategies clearly provide a broader, more comprehensive grasp of the "material world," so understood. However, this is not a sign of general cognitive superiority over competing forms of understanding, for it is a "world" (Chapter 7) not fully cohabited by proponents of competing forms of understanding, which are associated with stances towards material objects in which control is subordinated to other social values. The material possibilities of things do not exhaust the possibilities deemed valuable in these other "worlds" and many of them themselves lack value in them.

When one speaks of the "material world" without elaboration it suggests something that is shared and with which, regardless of value commitments, we all interact as a matter of course. Then understanding of it, the most encompassing possible, seems to make a claim on us all, as if it could be a sort of constant in our practical dealings with the world. When one speaks of the "material world" as the totality of the material possibilities of things, in contrast, the suggestion of universal interest retreats into the background. It is not that, for reasons connected with particular interests, the genuineness of possibilities encapsulated in theories consolidated under the materialist strategies is denied, but that the value of some (many) of them may be denied and so the interest in furthering their encapsulation is diminished. A competing form of understanding may wish to identify a class of possibilities that only intersects with that of the material possibilities, for example, those possibilities of things consistent with and supportive of ecological stability and a particular conception of social justice. Then, whether one opts to participate in developing materialist understanding or the alternative form of understanding cannot be grounded in the kind of comprehensiveness claimed for materialist understanding for it does not encompass the alternative class of possibilities. In valuing this alternative class of possibilities above the material possibilities and pursuing research to identify them, one does not therefore deny a claim made on us all that properly follows from our sharing an independent "material world."[12] Similarly, in ranking the cognitive value "wide-ranging explanatory power" above "full explanatory power" (or *vice versa*) one does not draw upon the general features of an object of inquiry shared by us all, but on the values that interact in mutually reinforcing ways with one's adopted strategies.

Taylor also refers to materialist understanding providing "recipes for more effective practice," increased "ability to make our way about and effect our purposes," increased "practical ability" and "increased practical capacity" (Taylor 1995: 48). Perhaps, then, we might think of the "material world" as the world as it must be grasped for the sake of most effective practical activity. But we cannot separate what counts as effective practical activity from the social world in which it is conducted and the conditions made available in it. Materialist understanding does not, in general as distinct from in a subordinate role, serve practical activity especially well in a "world" which aspires to attunement, or in one in which ecological stability and certain ideals of social justice are high aspirations. It does not enhance practical activity in general, only practical activity linked centrally with the stance of control of nature.

Technological (and experimental) success bears testimony that the world is amenable to grasp under the materialist strategies. As long as theories with wide-ranging explanatory power continue to be consolidated under the materialist strategies, technological innovations are assured (provided that the relevant social conditions remain in place), as also is gaining understanding of numerous domains of phenomena in which human causal influence is not relevantly present. Theories consolidated under the materialist strategies provide wide-ranging understanding of an extraordinary array of domains. But for many of the objects in these domains, including technological and experimental objects, it does not provide full understanding. We cannot, as it were, form a totality of all the objects in all the domains that materialist understanding ranges over and call this totality the "material world," for many of these objects play a role in human experience and practice, even though under appropriately formulated boundary conditions their behavior is well grasped in materialist terms. Concerning these objects, to engage in research aiming to gain wide-ranging rather than full understanding (or *vice versa*) does not rest upon purely cognitive grounds.

We do not find in comprehensiveness, therefore, a cognitive factor that can take us beyond displacement. As argued on p.135–6, materialist understanding properly displaces earlier forms of understanding because it, and not they, can grasp features of objects (particularly technological objects) that have become central in the "world" of daily life and experience. Displacement alone, however, does not provide, except perhaps by default (Chapter 7), a ground for the virtually exclusive adoption of the materialist strategies. Comprehensiveness does not provide it either, since materialist understanding is more "comprehensive" only in the sense of "wide-ranging," and cognitive interests do not suffice to prioritize wide-ranging to all versions of full understanding.

My earlier conclusion remains intact: it is their mutually reinforcing interaction with the modern values of control that explains the virtually universal adoption of the materialist strategies within the scientific community. We may say that they are adopted for the sake of grasping the "material world." But, if I am right, the most viable sense of "the material world" is that of the world as grasped under the materialist strategies, a "world" not consisting of objects that are ontologically independent of human observers, but partly constituted in practices that have mutually reinforcing interactions with the modern values of control. In this way I remove from Taylor's idiom traces of the suggestion that materialist understanding is generally (as distinct from selectively) applicable across perspectives of engagement, and that its pursuit has a purely cognitively grounded role in all cultures and epochs. My view is sufficient to account for technological success (and it recognizes that the material possibilities of things can continue to be grasped in an apparently limitless way), and it is unencumbered by the gratuitous universality that comes with taking "the material world" at face value, where the apparent absoluteness of the terms in the materialist lexicon disguises the link with the modern values of control. It also leaves open that adopting values that clash with the modern values of control may support the adoption of strategies in research which may provide bounds to the realm of materialist research that is considered worthwhile, or which may relocate the place of materialist research subordinating it to where it is relevant to a form of full inquiry.

My argument has left open that there may be alternatives to the materialist strategies, and it has indicated how commitment to values that challenge the modern values of control would render intelligible the quest to develop such strategies. But it has been objected that there really are no alternative strategies to be explored (McMullin 1999). Is this so? Certainly there is no alternative today institutionalized as materialist inquiry is and developed with comparable sophistication, systematicity, general credibility and power. Clearly, moreover, if there are no alternatives, the thrust of my argument is very much weakened. If alternatives cannot be identified, much of the force and social salience of the mutually reinforcing interaction between research under the materialist strategy and the modern values of control is diminished. If there are no alternatives then, although the mutually reinforcing relationship remains intact, the question about adopting the materialist strategies rather than some others is simply moot. But, if there are no alternatives, why is this so? I have already expressed doubts about arguments rooted in materialist metaphysics. Could it be that prevailing social forces do not permit or foster alternatives?

My suggestion is that in contemporary consciousness the categories of scientific materialism, reinforced by the hegemony of the modern values of control, have become so dominant (functioning as a virtual a priori) that the categories in which alternatives could be articulated have been marginalized or deemed pertinent only to expressive rather than investigative activities. Following this suggestion we might look for alternatives (or perhaps anticipations or residues of them) at the margins of the advanced industrial world or among minority approaches in mainstream institutions. That is why (Chapter 8) I will address efforts in some third world countries to integrate materialist knowledge and traditional local knowledge, and (Chapter 9) discuss alternatives that have been proposed from feminist perspectives. I do not look for *one*, but a diversity of more modest, less encompassing alternatives, complementing each other and materialist inquiry, reflecting the diversity of social and cultural values which have been adopted in different places. These are not alternatives in which social values will trump cognitive values, but ones in which social, cultural and material factors are considered together in complex, interacting causal networks (in projects of systematic empirical inquiry, whose products are subject to appraisal in view of the cognitive values), and in which the quest for those possibilities that may further human flourishing is foremost. The arguments developed in Chapters 8 and 9 will further illustrate how social and cognitive values play their roles at different logical moments.

Before proceeding to consider anticipatory alternatives, in the next chapter I will consider, in the course of a critical reflection on Kuhn, how social and cognitive values played their respective roles in the process of the transition from pre-Galilean to Galilean science. I will rebut Kuhn's account (one that, unlike Taylor's, does not rest upon seeing technological success as a symptom of cognitive superiority) of how the cognitive values alone account for the rationality of this transition, and in doing so reinforce my conclusion about the grounding of the materialist strategies in the link with the modern values of control. This will further open up the space for entertaining alternatives to the materialist strategies.

7 Kuhn: scientific activity in different 'worlds'

I have maintained that social and cognitive values both play important roles in scientific activity, but the roles are played at different logical moments. Social values can have a legitimate role concerning the adoption of a strategy, but a theory is properly accepted of a domain of phenomena only if it manifests the cognitive values to a high degree. Consistent with this there may be alternative strategies (to which the materialist strategies are subordinate or with which they are incompatible) under which theories, manifesting the cognitive values to a high degree, could be developed. In this chapter – looking for insight into how to make sound theory choices when conflict extends across strategies, and thus laying the groundwork for a defensible account of *impartiality* in Chapter 10 – I will reflect on the historical moment at which the materialist strategies began to come into their dominant position, prior to which (in Europe) the strategies of Aristotelian science largely framed scientific activity. At this moment, there were alternative strategies in competition, and I will argue that social values actually played key roles in supporting the adoption of the respective strategies, and a crucial role in bringing about the virtually uncontested adoption of the materialist strategies (further consolidating the conclusion of the previous chapter). Here I disagree with Kuhn, who maintains that the Galilean (materialist) strategies were adopted principally because of their demonstrably greater fruitfulness in generating theories that manifested the cognitive values to a high degree.

KUHN'S "NEW-WORLD PROBLEM"

With his famous aphorism: " … though the world does not change with a change of paradigms, the scientist afterwards works in a different world" (Kuhn 1970: 121), Kuhn points to an important feature of transitions, like the one from Aristotelian to Galilean science. In what sense did Aristotelian scientists work in a different 'world' from the one Galilean scientists work in? How do these 'worlds' exclude each other? Why work in one 'world' rather than another, and when is the scientist confronted with that choice? Are there cognitively (rationally) compelling grounds to work in the Galilean 'world?' What changes, and what does not change, with a fundamental change in the strategies that frame research activity?[1] Can paradox be avoided if one accepts Kuhn's further claim that scientists, working in different 'worlds,' observe different (and incompatible) things when looking at the same object? It is a cluster of questions like these that "the new-world problem" (a term coined by Hacking 1993; see also Kuhn 1993) designates.

The world, scientific 'worlds,' social "worlds"

There is only one world, the repository of all possibilities, or the totality of things, events and phenomena that constitute the causal order of which human activities and experiences are a part. *The world*, characterized abstractly in this way, can be thought of, for example, as the Aristotelian cosmos (shaped by teleological causal principles) or the modern spatio-temporal totality (structured by quantitative laws). *The world* that "does not change with a change of paradigms" is, for Kuhn (I take it), the causal principles and fundamental constituents of the world. Both Aristotelians and Galileans agree that they do not change, though they differ about what they are.

Any talk of "worlds," therefore, is essentially metaphorical and, in colloquial idiom, it tends to move among multiple layers of meaning. Following the metaphor, a "world" is a kind of self-contained totality, as it is grasped, interacted with and articulated by its "inhabitants"; unlike *the world*, a "world" does not exist apart from the practices, modes of interaction, self-understandings, and articulations of its human inhabitants. In colloquial usage, "worlds" may overlap and be contained in other "worlds," so that their limits and dependencies, and that there are alternatives to what is taken for granted in them, may not be recognizable as such from within; and they all manifest historicity. The self-containment and all-inclusiveness is only more or less, allowing that a "world" may be susceptible to "outside influences" (coming from *the world* or other "worlds"), which sometimes explain important happenings in it, but which the categories deployed in its commonly shared articulations may be ill-suited to grasp. The historical sustainability of a "world" depends upon its being able to gain or to maintain its high degree of self-containment and all-inclusiveness in the face of conflict and competition for resources with other "worlds." Ultimately, then, to understand well what happens in a "world," one needs to gain access to an explanatory scheme that transcends the provinciality of one "world" and to be able to compare "worlds," recognizing how they interact with one another in an ever more encompassing movement, so that one will recognize layers of "sub-worlds" contained in larger "worlds." It is doubtful that there could be a "world" that contains all "worlds"; but the larger a "world" is the more relatively self-contained it is likely to be.[2]

What Kuhn means by "the world in which the scientist works" has been interpreted in many different ways, and what truth, if any, it may be attempting to disentangle about scientific change remains a matter of controversy. Some (for example, Sankey 1994, 1997), who consider semantical incommensurability to be an important (though limited) phenomenon, tend to dismiss the aphorism as simply inflated rhetoric on the ground that it ignores that different theories may refer to the same "mind-independent objects" even though their respective referring terms are not inter-translatable. Others (for example, Hoyningen-Huene 1993) have provided a neo-Kantian interpretation, identifying scientific 'worlds' with "phenomenal worlds," and *the world* with the epistemologically inaccessible "world-in-itself." Still others (for example, Rouse 1987) emphasize that a 'world' is linked with a "form of life," its required skills (habits, expectations and sense of what is possible), its organizing structures, and its ways of actively engaging in research[3]

Developing the last line of interpretation, I take a scientific 'world' to be constituted in practices that are shared among a community of scientists – data gaining, theory formation and appraisal, and theory application practices – and in the shared beliefs and categories deployed to make the practices, their conditions and their outcomes intelligible, communicable and effective. A 'world' is the set of objects that becomes articulated in and by the characteristic practices shared within this community, and that are interacted with (in characteristic modes) and whose possibilities are investigated (and sometimes brought to realization) in these practices.

According to Kuhn, scientific activity is necessarily carried out within a 'world,' and any understanding we have of *the world* is always gained within a 'world,' for we can only investigate *the world* by investigating objects that are appropriately characterized and appropriately interacted with. Within a 'world,' theory developing and data gaining practices are reciprocally related by strategies that they deploy: strategies that constrain the kind of theory that may be entertained and lead to the selection and seeking-out of empirical data with certain characteristics – such as the strategy to entertain only theories that can be constructed with the specialized vocabulary of a chosen lexicon, and to select data that are expressed in categories appropriate for being put into relationship with such theories for the sake of appraising how well they manifest the cognitive values. We might say that the object of scientific inquiry (or an approach to scientific inquiry, Chapter 5) is that which can be grasped under the adopted strategies.

Historically there have been changes of 'worlds' in which scientific activity is carried out. A change of strategies accompanies (and partly causes) a transition from theory to theory – like the one from Aristotelian to Galilean theory – where the theories on either side of the transition are incommensurable: the lexicons in which they are respectively formulated contain key terms that cannot be intertranslated; and empirical data, expressed without implicating one or other of the lexicons, generally are not available as a decisive ground for choice among the fundamental theories. The historical reality of these transitions challenges – and Kuhn's aphorism suggests how deep the challenge is – the common (logical empiricist) accounts that transitions from theory to theory can be explained, and ideally shown to be rational, in terms of relations between the theories and a body of empirical data that all of the competing theories can be expected to encompass.

The challenge requires that the question of the rationality of theory choice be relocated so that it is linked with that of the rationality of adoption of a strategy. Thence the new-world problem. I will give special attention (later in the chapter) to the question: Why (or why not) engage in the scientific practices which deploy a new strategy? This question cannot justifiably be reduced to one about which fundamental theory to accept, or to any questions about soundly held beliefs. Change of strategy does involve change of lexicon (whose expressive power defines the limits of the possible of the new 'world'), and thus of the categories with which admissible theories are formulated. This, in turn, may involve change in general beliefs about the object of inquiry, its general features, and fundamental causal principles, as well as of the theories that are accepted. But the new strategy is not adopted as a rational response to the new beliefs, for the beliefs, theories, and strategy develop together in complex, on-going interaction within the new and developing patterns of scientific activity. (That is why, after the transition is complete, the gulf between the new and old 'worlds' can be so vast.) Even if the general beliefs about the object of inquiry and fundamental causal principles are considered as (metaphysical) beliefs about *the world*, at least for a period of time the new strategy cannot be rationally grounded as the one especially suited for generating concrete knowledge of *the world* (since, pending the unfolding of the strategy, that the new metaphysical beliefs about *the world* constitute knowledge, could not be rationally grounded). Consequently, at least during this period, support for adopting the strategy and for accepting any accompanying metaphysical beliefs, must partly come from other sources. One source might be the (social or moral) value attributed to the mode of interaction with things which is prioritized in the new scientific practices. In that case, the source would be in the social "world" in which the competing scientific 'worlds' are located.

Kuhn's discussion of the question about the adoption of a strategy concerns only *the world* and scientific 'worlds,' but not social "worlds." But a scientific 'world' is always part of a social "world"; scientific practices have social conditions, and scientific theory is applied in the prevailing social "world"; and changes of scientific 'worlds' and social "worlds" often accompany each other. Bringing social "worlds" into the picture, we will see, helps to relieve some of the paradox that may seem to threaten Kuhn's position.

The new-world problem incorporates also "the old-world problem": What are we to say about the old scientific 'world' and, above all, about the objects that are said to constitute it and the kinds of objects said to be observed in it? Aristotelian theory contains posits about natural motions (motions directed towards natural ends) and the motions of the heavenly spheres. Are we to say that these objects existed in the Aristotelian 'world,' but no longer exist in ours? Or, are we to say that these objects do not, and never did, exist? The first alternative has the ring of absurdity. But, given the incommensurability of Aristotelian and our science, how could we rationally justify opting for the second?

Kuhn (1970) describes the theory transitions under discussion as revolutionary, making an analogy with revolutionary historical transitions between incompatible political orders or political "worlds," such as absolute monarchic and democratic "worlds." Does this analogy provide guidance for the old-world problem? It is clear enough what we are to say about the old political "world": it once existed, and now it does not for it was overthrown and replaced by the new "world"; monarchic institutions and the roles they nurtured once were real, and now they are not. Moreover, we are comfortable in explaining political revolutionary transitions largely in terms of such factors as material and social conditions, personal and social values, and power. From the perspective of a social science theory of the new democratic "world" we may be able to explain the incompatibility of democratic and monarchic "worlds" and weaknesses in the monarchic "world" that may not have been apparent from within, we may reject beliefs that were essential to the maintenance of monarchy, we may value monarchic institutions differently, but we will not deny the existence of these institutions and their constitutive roles.

Kuhn, unlike Feyerabend (1989), shrinks back from embracing the radical implications of following the political analogy wholeheartedly. In the Aristotelian 'world,' as portrayed in its theories, there are natural motions and the like, and the practices (including observational ones) of Aristotelian science are characterized by its practitioners in terms of interactions with and relations to such objects. In contrast, in the Galilean 'world' there are lawful motions (motions that fit differential equations, undirected to ends). If the political analogy carried through in detail, we might be led to conclude: once there were natural motions, now there are not; now there are only lawful motions. It also might suggest that, as with monarchic institutions, we have excluded natural motions from our 'world' for reasons of social value or socio-economic power, and thus that the theory transitions are explicable primarily in terms of social factors. Here Kuhn makes the crucial break with the analogy: the theory transitions can be explained and justified in cognitive (epistemic or rational) terms. Because of that we properly conclude that natural motions do not and never did exist (see pp. 168–70); Aristotelian scientists believed that they did exist and, given their epistemological horizon, their beliefs were well-founded, but they were mistaken. Across the gulf of the theory transition, not only are the old and new theories incommensurable and products of incompatible practices, they also are cognitively incompatible. In some sense both the old and new theories are (in part) about the same things (objects in *the world*?), and (in some sense) in some respects the new strategy enables us to grasp certain phenomena and spaces more adequately. It is not that we have excluded natural motions from our 'world' because we deny a certain kind of value to them; we deny that *the world* contains them. There is no counterpart of this in the political analogue.

Kuhn's metaphor of 'worlds' now appears as a metaphor of a metaphor!

As 'worlds' problems multiply, one might wonder if these problems are artifacts of using the metaphor, rather than problems driven by the phenomenology of scientific change. Why not, one might ask, simply say that the Aristotelians were fundamentally mistaken about *the world*? Why say that they conducted their scientific activity in a different 'world' when central objects with which they attempted to organize and explain their experience do not exist? They said that they were investigating the properties of natural motions, but they were not doing that for there are no natural motions to investigate. So they were radically misdescribing what they were doing and what they were observing. We and they both investigate *the one world*, and we have evidence that our theories offer better accounts of it. This is clear enough. Why confuse the matter by piling metaphor upon metaphor?

Kuhn resists the impact of this line of questioning, in part because he does not think that it captures the magnitude and (above all) the range of dimensions of the transition. It is also in part because he rejects what is often an undercurrent, that we (modern scientists) have got the broad lines of the causal principles and fundamental constituents of *the world* right, so that we should not be considered to be doing science in a 'world' of comparable status to the Aristotelian 'world.' For Kuhn, as pointed out on p. 150, any understanding of *the world* is gained within a 'world,' whose strategies (and thus the character of the practices in which they are deployed) partly constitute the object of scientific inquiry. In this (and only this) respect Galilean and Aristotelian science are on an equal footing. From our viewpoint, we can affirm justifiably that certain objects of Aristotelian theory do not exist in *the world*, and that the Aristotelian lexicon reflected more the character of the practices of Aristotelian science and the values linked with them than features of *the world*. But our viewpoint does not ground that the Galilean lexicon reflects features of *the world* more than the character of the practices of Galilean science, and the values linked with them. We have no ironclad assurance that our 'world' will not change, even go the way of the Aristotelian 'world,'[4] and that our strategies might not come to be supplanted too. There is nothing about scientific practices that entitles us to affirm that theory offers understanding of *the world*, or objects in *it*, independent of their relations with human beings. That doesn't mean that we don't gain knowledge of *the world*; we are part of *the world*. We grasp *it* against the background of an essentially historical, structured lexicon. That is our 'world,' the world of which we have knowledge, a 'world' constructed in the course of scientific practices.

What is clear is that our strategies are working very well in solving puzzles today, just as (though with much greater effectiveness than) the Aristotelian one did for a time in its own way in its own 'world' before (according to Kuhn) dissolving in the face of internal crisis and the emergence of a more compelling competitor. Whatever its weaknesses may be diagnosed to be from the viewpoint of the Galilean 'world,' it remains that, in its own time, Aristotelian theory provided of a significant body of experience what were reasonably taken to be illuminating explanations that were superior to available competitors. In order that doubts could be formulated in a probing way about whether its core categories referred to things and kinds in *the world*, a new kind of competing 'world' needed to be in play. A new kind of competitor, deploying a new strategy, could sow similar doubts about our current lexicon, though it would leave intact (even if no longer relevant for theoretical or practical ends – cf. the discussion of alternative strategies in Chapters 8 and 9) a considerable body of empirical knowledge. It is fair to say that our best theories, insofar as they apply to certain specified domains, have been submitted to far more severe testing than their predecessors.

OBSERVATION IN GALILEAN AND
ARISTOTELIAN 'WORLDS'

According to Kuhn, a "good" theory, regardless of the strategies under which it is developed, is related to the relevant empirical data, and to other theories, in such a way that the cognitive values are manifested to a high degree (Chapter 3). The empirical data – the observed facts – are beliefs gained properly through acts of observation and interactions of humans with objects which fix beliefs. Beliefs reflect intentionality. They are related with other beliefs with which they exhibit logical, rational and evidential relations, and as such can be expressed as having propositional content (Chapter 3). Beliefs, thus, presuppose a lexicon. For Kuhn, there is no "observational lexicon" or special vocabulary that does not draw from theoretical categories that is especially apt for reporting the observed facts relevant to evaluating scientific theories. I will now illustrate this by showing in detail the way in which much of the empirical data, relevant within scientific activity, bear the mark of strategies, and thus cannot serve generally to resolve issues of theory choice that cut across the strategies. This detail will throw light on how 'worlds' are incompatible and will point towards solutions of the new-world problem.

Compare formulations of physical theories of Aristotelian (A) and Galilean (G) types.[5] On the one hand, they aim to express understanding of objects of *the world*. On the other hand, they express different ontologies, and refer to different natural kinds. The intentional objects of A and G are different. A is about such things as the natural motions of things and the terrestrial elements, fully characterized in qualitative terms; whereas G is about the underlying law, structure and process of things, characterized in mathematical terms. Neither theory has the lexical resources to express in its own terms the full ontology of the other. This fits with the metaphor: A and G are about objects in 'different worlds.' And the differences go all the way down to the observational data. Exploring the implications of his aphorism, Kuhn added: "when Aristotle and Galileo looked at swinging stones, the first saw constrained fall, the second a pendulum. ... A pendulum is not a falling stone ... Galileo saw the swinging stone differently" (Kuhn 1970: 121–3). Different things are observed in the two 'worlds.' Yet A and G make their observations while "looking at" the same thing:

> the proponents of competing paradigms practice their trades in differ-
> ent worlds. One contains constrained bodies that fall slowly, the other
> pendulums that repeat their motions again and again. ... Practicing in
> different worlds, the two groups of scientists see different things when

they look from the same point in the same direction. ... this is not to say that they can see anything they please. Both are looking at the world, and what they look at has not changed. But in some areas they see different things, and they see them in different relations one to the other.

(Kuhn 1970: 150)

Although the theories of A and G are incommensurable to a significant extent, their theoretical categories figure respectively in observational reports made while looking at objects in the one world. But, in the lexicons of A and G there is no way to state that the reports are about the same thing, for the referring terms (inhibited free fall and damped pendulum) are as different as the referring terms used to describe the institutions of two societies separated by a revolutionary divide. And there are no observational reports that do not use theoretical terms from one or other of the theories that both A and G would *not only* assent to, *but also* deem relevant to their theoretical deliberations.

A and G are *different*. They are *incompatible* only if, at least in part, they are about the same objects. To settle whether they are about the same objects, we need to go beyond the conceptual resources of the two lexicons, to go beyond inter- and intra-theoretical reflection and, I will argue, pay attention to the role of theory in *application*, for instance in providing understanding of phenomena in the realm of daily life and experience, and guidance in carrying out practical activities.[6] These phenomena may be observed within common practices in which both of the competing theorists participate in their daily lives. Consider a child's swing.

Observing things from the perspectives of different practices

There are countless (logically) distinct observed facts about any thing. G and A observe many of the same facts about the swing: that it is a swing, that it has a certain spatio-temporal location, that its chains have a certain length, that it needs painting. Which observed fact about a thing is considered at a given time depends on context, purpose, audience, and activity: playing with a child, giving instructions to a repair-person, engaging in scientific activity. Being able to identify the same thing across practices and contexts, and so being able to refer to the same thing using expressions with different meanings, is a normal part of learning the meaning of basic referring expressions. It is one and the same swing that my child is playing on, that I bought in the store, that was repaired yesterday – and to whose movements I am trying to apply the scientific theories I endorse.

Outside of scientific activity, G and A agree on most of the observed facts. That is how we know that they are looking at the same thing, and that what they are looking at does not change with the change of strategy. There are social practices in which there is interaction with objects to which the theories on either side of the transition are both applied and that do not change with the theory transitions. Using their vocabulary, objects looked at by both theorists can be described in a way to which they both assent. That is enough to assure that A and G, regardless of the differences in what they observe when engaged in scientific activity, are looking at the same thing; and it consistent with Kuhn's view that we cannot grasp *the world*, except from within a "world." That *"the world* does not change" plays no operational role in providing this assurance. *Within* their respective scientific activities, A and G report different observed facts about the swing when they look at it – *different* and also *incompatible*.

At one level, the difference is like that present when a coach, player, commentator and spectator each describes a particular episode in a soccer game. They observe different things because of their different perspectives, modes of engagement, interests, available vocabularies, and the capabilities to make the discriminations necessary to deploy the relevant vocabulary. But these descriptions need not be incompatible; and, if they are, with "correction" compatibility can readily be obtained. On the other hand, A's and G's observational reports are not incompatible in the way in which reports of a thing that it is completely red and completely green are incompatible, where at least one is straight-forwardly erroneous, and whose resolution requires, for example, further observation and careful attention to what is being observed. I will return to the question of what sort of incompatibility is involved.

Observation and its goals within an explanatory practice

Let us consider an observation to be an action of looking at a phenomenon with the immediate goal to produce a belief (expressed in an "observational report") that is appropriate to the activity being engaged in. Then, the immediate goal of observation is subordinate to the objectives of the activity at hand.

When the activity is science, the objective may be, for example, relating the phenomenon to the causal order in which it is produced and has consequences, or encapsulating the possibilities open to it (Chapter 5). Then, categories will be used in scientific observational reports that enable them to be connected with proposals about the causal order, and so they may be drawn from the specialized vocabulary of one's adopted lexicon. Moreover, which observations are made may vary with views (that may be implicit in the lexicon) about the fundamental causal order and about the best available general theories. These, in turn, are connected with differences in scientific practices, including disciplined, shared and stable norms of observation. What a phenomenon is observed to be derives (in part) from practices and presuppositions, and involves its place in the practices.[7] What is observed – the phenomenon, under the description of the observational report – is constituted (in part) by human intentionality (McDowell 1994), and it is a product of an interaction: *both* a way of looking at the phenomenon or at it under certain conditions or at certain sequences of its states, *and* looking at it in relation (a product of our interactions) to certain objects. A observes that the swing (in motion) is in the state of inhibited free fall; G that it is a damped pendulum.

The temporal interval of an observation

Any observation takes place over a period of time, so that what is observed may vary with the sequence of events held to be constitutive of the phenomenon being looked at. For A, in our example, the relevant sequence ends when the swing comes to rest as close as it can get to the earth's surface. That it is observed to come to rest there is critical for linking the phenomenon to the general causal order, as represented in A's physical theory. For G, the relevant sequence is much shorter, including a sufficient number of the back and forth movements to enable the period to be measured (Kuhn 1970: 119, 123).

Both A and G, no doubt, would assent to the following two observational reports: that the swing eventually comes to rest, and that its movements have a period of approximately (say) two seconds; just as they would both recognize such other facts as the color of the swing, the rusting of its chains, and the sounds of pleasure coming from the children playing on it. Scientific theory is not expected to "fit" all observed facts, only the ones appropriately selected in the light of scientific objectives. A and G agree that none of these other facts are pertinent for the objectives of scientific practice; but they disagree on the pertinence of the first two reports. For A, the periodicity of the swing's motion has little more relevance than the rust on the chains; for G, that the swing comes to rest is a fact to be explained, but not one critical for grasping the causal order of things.

Each of A and G looks at the sequence he deems relevant and draws from the categories of his adopted lexicon; then A observes "inhibited free fall," and G "damped pendulum." Within their relevant communities, such observations are routine, spontaneous, stable, marked by intersubjective agreement, and not inferred from reports made in a purely "observational language"; and the phenomenon is replicable. In the present case, A and G both could have observed differently; they could have produced observational reports to which both would assent, and had they been engaged in another activity together (for example, consulting with a repair-person) no doubt they would have. When pursuing scientific objectives, however, characteristically they do not.

One might query: where we confront two stable but incompatible observations of the same phenomenon, should we not restrict the relevant lexicon for observational reports to that which is shared among the disputants? Then, would it not be clear that the disputing parties are simply looking at different sequences of events, each of which is now described in a way that all can accept? Is not G's sequence simply contained in A's, with A and G attending to different aspects of it: A that the swing come to rest; G that the movements are approximately periodic? Then, should not the theories of both be judged in terms of their "fit" with the common set of data?

Observation and what is looked at; observation and measurement

In fact, however, G does not just look at some of the movements in a sequence, while A looks at all of them. A may not look carefully at the back and forth movements at all. More importantly, G looks at the *relationships* of the movements to various instruments, such as clocks. G *measures* rather than simply *observes*. G's and A's "less theory-laden" observed facts are, to a large extent, distinct. They need not be totally distinct. G and A compete. G, too, observes that the swing comes to rest, and he offers an explanation of it; but, that it comes to rest, is irrelevant as an explanatory factor of the motion. They both look at the swing. But that underdescribes what each looks at directly, and misses the differences in the ways they look at it. G's way of looking at the swing is to observe its relationship with instruments, which he himself makes part of the phenomenon. G, through measurement, instrumental and experimental interventions "creates" many of the phenomena that he looks at (Hacking 1983); A looks at "natural" phenomena. For A, G's observations are of "artifacts," not of "nature"; investigating them can illuminate human objectives, but not the natural order. For G, A's reports cannot gain us access to the causal order of the world, which is marked by quantity and lawful relations among quantities.

As indicated already, A may assent to G's (less theory-laden) observed facts (for example, that the period is two seconds) while regarding them as irrelevant to the objectives of scientific practice. Although such facts may elude the predictive and explanatory scope of A's theory (like the rusting of the swing), they do not contradict it. With the developing sophistication of G's practices, however, his observed facts may only be recognizable by someone skilled in the relevant instrumental interventions and experimental manipulations, much as an unskilled spectator (or a skilled rugby coach) cannot observe all that a soccer coach observes in the course of a soccer game. The phenomena they describe come into being only in the course of a practice involving such skills (Rouse 1987).[8] In these cases, G's observed facts make no claim that A, given his adopted strategy, need take into account. A has no reason to consider the testimony of G authoritative for contexts outside of those of G's practices, for G's practices concern another realm of possibilities, not those of nature; just as a soccer coach has no reason to consider the testimony of a rugby coach authoritative outside the context of rugby. A and G are "playing different games." Like soccer and rugby, they cannot be played on the same field at the same time (Taylor 1982) – but which playing field is the swing in?

Observational reports and how they are theory-laden

Any observational report involves classification, representing the phenomenon as of a kind, and thus treating it as like certain other phenomena. skillfull observers, in the act of observing, make the relevant classifications spontaneously without engaging in inference. In observing the swing, the phenomena that A and G (both skillfull observers) consider the swing's motion to be like are different.

A relates the swing's motion to that of freely falling bodies, where "relate" means first to affirm a certain likeness, and second to imply that it is deviation from free fall that needs to be explained. In science observations may be made for the sake of displaying a phenomenon as part of a causal order; then, we observe in a way that facilitates that objective. A relates the motion of the swing to a "natural" motion. In contrast, G relates it to the motion of a pendulum without retarding forces. For G, it is departures from the pendulum's motion (represented ideally) that need to be explained. To understand the swing's motion, G relates it to a phenomenon which is an experimental artifact, a product of skillfull human activity, more exactly, to an idealization of this phenomenon.

In this way each observational report is theory-laden, for instance it presupposes a theory and deploys its categories, respectively A's theory of natural motions and G's of the pendulum. For A, G's mode of relating is doubly problematic: such pendula do not exist, but rather they are idealizations; and even as approximations, they do not exist "in nature," but only as constructed in experimental spaces, as human artifacts.

G's observation, thus, presupposes working in a certain sort of 'world,' one in which the practices of skillfull experimentation and measurement, and mathematical idealization have been developed among a community of scientists. A theory of the pendulum's motion is grounded in phenomena, aspects of which are measured, in experimental spaces whose boundary conditions, as well as the initial conditions of sequences of events in them, are humanly controlled. Thus, it is grounded in observed facts of phenomena that are the causal consequence of planned human intervention. Again, given A's outlook, while he may assent to many of these facts, they have no relevance to (his notion of) the objective of science, and the lexicon upon which his theories draw contains no categories which could usefully redescribe these facts. They may be relevant to a skilled artisan's practice (dealing with, for example, clocks, musical instruments, pumps or weapons), but not to science.

ARISTOTELIAN AND GALILEAN OBSERVED FACTS: DIFFERENT AND INCOMPATIBLE

A's and G's data gaining and theory development practices are significantly different, involving different immediate objectives, skills and patterns of observational discrimination, so different that their core lexicons cannot be inter-translated. Are they just different "games" – each with its own lore, styles, skills, possibilities of achievement, rules and mode of discourse – that, like soccer and rugby, could coexist in the same social "world" at the same time (though not on the "same field")?

We sense, however, that there is not just difference, but deep incompatibility. It seems that the only way for A and G to coexist (in the long run) would be if they concerned completely distinct domains of phenomena. But, despite the differences discussed in the previous section, each offers an explanatory account of the motion of the swing and of other significant phenomena in the realm of daily life and experience, for instance of those things, events and phenomena that are confronted commonly in daily life in the social "world" in which the scientific 'worlds' are located, and which structures such activities as those directed to meeting basic needs and to the ends of economic production. How are A's and G's observations of the swing incompatible? Kuhn is right that it is not in virtue of inconsistency expressed within the lexicons deployed by A and/or G. Could the observations be considered complementary, as the descriptions of the swing as a child's plaything and as a commodity for sale in a store are?

Practical and cognitive aspects of incompatibility

Two features seem to mark the incompatibility. One is practical or pragmatic. It is that the scientific practices (like those of soccer and rugby), including the data gaining practices, of A and G are incompatible: one cannot simultaneously observe a natural (in A's sense) phenomenon and measure it or observe it under human intervention; one cannot simultaneously relate a phenomenon to its place in the cosmos and relate it to an idealized model of an experimental phenomenon.

The second feature ensures cognitive, and not just practical, incompatibility. It arises when we apply either of the theories to the swing's motion. Then, we make use of vocabularies that extend beyond those of the two theoretical lexicons, and in effect include items from the lexicons among the categories of daily practical activity. We thereby gain greater expressive resources with which we can derive a contradiction from the two observational reports: from "the swing is in a state of inhibited free fall" we infer "it is moving towards its natural end, the center of the universe"; from "its movements are those of a damped pendulum" we infer "it is moving in accord with mathematical laws" – but, if it is moving in accord with mathematical laws it is not in a state of natural motion, and thence a contradiction is derived.[9] To derive the contradiction we need the greater expressive resources of the lexicons of daily practical activity, which can serve as a metalanguage of both theories. A theoretical lexicon can be looked at from two perspectives. On the one hand, it is learned and the skills for its deployment are gained during an apprenticeship, in the course of practical engagement, using its categories in connection with observational (measurement, experimental) and theoretical tasks. One simply learns the lexicon in use. On the other hand, it can be discussed in a metalanguage, compared with other technical lexicons, and the reasons for its adoption investigated. According to Kuhn, the former can happen without the latter and most of the time ("normal science") it does. As long as the lexicon remains a fruitful means for resolving puzzles so that its comparison with other lexicons is not an immediate issue, this need cause no tension. Then, contradiction that can only be formulated at the metalinguistic level remains invisible, and to practitioners who have not become fluent in the metalanguage the difference between A and G would appear to be similar to that between soccer and rugby, having as much impact on their consciousness as cricket has in the USA.

These two features of incompatability are not specific to the case of the swing; they have systematic sources. Neither A nor G makes an observational "error" in observing the swing – as G does, for example when he mismeasures the period of the swing, or A when he reports rising smoke as an instance of inhibited free fall. Such errors can easily be corrected without disruption of theory. The "error" in the present case resides in the acceptance of a theory and ultimately in the adoption of a strategy (and its associated skills and practices). There is no relevant "correction" of either of the two observational reports that can be made without abandoning the strategy under which the observation was made. The incompatibility of the observations derives from a systematic incompatibility of the strategies of A and G, which has a cognitive (logical) dimension in that, as just argued, when embedded in an appropriate metalanguage, reporting observations with the categories of the respective theoretical lexicons leads to contradiction. It also has a pragmatic dimension: the data selection strategies are parts of (locally) incompatible practices. From the cognitive dimension of incompatibility, if follows that one cannot choose both of A's and G's theories of domains of phenomena that contain the swing's motion. But the practical dimension precludes empirical adequacy grounding a decisive choice between them.

Incompatibility pertaining to application

There is also a third dimension of incompatability, pertaining to the realm of daily life and experience: the competing theories cannot be applied together to objects in this realm, so that in the long run the broader social "world" cannot contain as "sub-worlds" both of the scientific 'worlds' constituted respectively by the practices of A and G (even, as it were, on "different fields").

Certain constancies in the realm of daily life and experience, maintained through the period of theory transition, ensure – I have argued – that (in part) A and G are about the same things, and thus incompatible. The swing cannot be both an instance of impeded fall and a damped pendulum, but it lends itself easily to either interpretation, so that its role in social practices remains largely unaffected by the outcome of the conflict between A and G. On the other hand, during this period the realm of daily life did change in significant ways, as the historical processes of shaping modernity were under way. The characteristic stance towards natural objects, as manifested in mechanical innovations and expressed in the leading articulations of the time, moved more to that of control and away from adaptation or attunement. Through this there came about a striking change in the objects that became central to the structuring of daily life, or at least to setting its transforming trajectories. Before the transition, A applied most easily to many of them; after it, to an increasing extent, G did. After the transition, the scientist worked not only in a new scientific 'world', but also in a new social "world."

A and G do not apply in the same way, and they cannot be applied together coherently. Upon development, they provide different anticipations of what is possible in the realm of daily life and experience, different accounts of the limits of the possible, different guidance about means to ends and the consequences predicted to follow from actions, and different explanations of how things work. Each theory, upon application, partly defines possibilities of a social order shaped by a particular conception of human well-being or flourishing, and a particular characteristic stance of human beings towards material things – A that of "adaptation", G that of "control" (Chapter 6). Thus, these differences may also be accompanied by different assessments of the significance (value, worth) of the kinds of possibilities anticipated, for instance different judgments about the value of realizing them in the realm of daily life.

The possibilities, that A and G respectively highlight, cannot be realized together systematically in any historical "world." A social "world" increasingly shaped by relations with mechanical objects (to which G easily applies) undermines the conditions of stability and relative constancy (where A easily applies) required to articulate coherently daily life and aspirations towards flourishing in terms of attunement to nature or to the cosmos; conversely, commitment to maintain such stability poses obstacles to projects of technological development.[10]

After the period of the theory transition G, but not A, was widely applicable, at least in the practices of the ascendent powers that were reshaping the social "world" in the direction of mechanical objects becoming of major salience in daily life and productive practices. *Before* it, A applied well within the "world" of medieval Christendom as this "world" was represented in its dominant articulations; whereas, even retrospectively and counterfactually, G would not be much applicable in this "world," except to phenomena that were relegated in the dominant articulations to the status of artifact rather than of nature.

A applied well to numerous objects of salience in the realm of daily life and experience, and its practices were cultivated, first in the historical, social "world" of ancient Greece and later in that of medieval Christendom. A's 'world' was a "sub-world" of these larger "worlds" or cultures, fitting neatly with them and the self-interpretations borne with them because of several mutually reinforcing features: the ubiquity of teleology (and meaning); emphasis (which became articulated theologically with the backing of the Church's authority and power) on hierarchical structure in the cosmos and in society, with the two structures considered as ordained by God to mirror each other (Chapter 4); attention to the many-sidedness of things and interest in full understanding (Chapter 5); the characteristic stance towards nature articulated in terms of notions like "attunement" linked with "contemplation" or "being in touch with the natural course and rhythms of things"; the sense (though not always the actuality, especially as the "world" of medieval Christendom entered into fatal crisis!) of a stable order, or at least that there were no significant or desirable new possibilities to be discovered; and a strong distinction between natural objects (responsive to natural ends) and artifacts (responsive to ends imposed by human beings).

No "world" is ever fully self-contained, all-inclusive and without ten-dencies that run counter to what is articulated in its reigning self-understandings, and these counter-tendencies can cause crisis in it and become the source of a reshaped social order. Immediately before the period of the theory transition, the "world" of Christendom was, of course, in crisis. Within it, projects linked with ascendent values and powers that were to displace it involved increasingly a place for mechanical objects (to the workings of which G, but not A, could be applied easily); and the more they found a place in the realm of daily life and became recognized as a key part of the transforming trajectory of the social order, the manifestation of the cognitive values in A of the key objects in the realm of daily life diminished. I reiterate that those possibilities of G that were realized at the time of the transition were (with few exceptions) not products of applied science in the contemporary sense, but objects of which G, but not A, could give an interesting explanatory account. What I am calling G's possibilities includes (extending far beyond those actually foreseen in the seventeenth century), on the one hand, those that can be grasped from successfully establishing theories that represent them lawfully and, on the other hand, those that are opened up through adopting the stance of control as the characteristic one towards material things. Those possibilities that fall under the second description in general also fall under the first (Chapter 6); but not always conversely (the planets).

As Kuhn has made us well aware, there were numerous anomalies confronting A, addressing some of which led to its late medieval modifications, including projectile and circular motions, free fall, quantitative treatment of motions and forces, the discord between Ptolemaic astronomy and Aristotelian physics. The anomalies were well known, pertinent to phenomena encountered in the realm of daily life and experience, not ignored, but typically dealt with either by introducing auxiliary (including *ad hoc*) hypotheses or by instrumentalist interpretations. Since G came to shed light on these phenomena, it is tempting to say, counterfactually, that G could have solved problems faced in the earlier social "worlds." But G could not have done so without creating discord in those "worlds" and their self-understanding (as it later did) – unless G's terrestrial science was interpreted as pertaining to the domain of artifacts only, and the Copernican theory interpreted instrumentally (as in Osiander's preface to Copernicus' *De Revolutionibus*). It could not have illuminated the characteristic modes of interaction with material objects and with other people in these worlds. In the prevailing self-understandings, the anomalies were marginal to these characteristic modes. Then, inquiry that focused on the objects encountered in the characteristic activities of these "worlds" would have no place for G's type of inquiry, since the latter is not a mode of inquiry that seeks full understanding, but rather abstracts the objects of inquiry from their social, value-related and ecological dimensions. It is not a matter of ignoring the evidence or shutting off alternative modes of inquiry, but of focusing inquiry on key objects of the realm of daily life and experience prevailing in the "worlds." Theories, even *our* best established ones, cannot be grafted onto any social order.

My suggestion is that the applicability of G (or of A) – both as offering accounts of important (in the self-understanding of a "world") objects in the realm of daily life and experience, and as providing knowledge that informs important social practices – depends on what are the characteristic activities of a social "world." One cannot enter the 'world' of G without exiting the "world" of medieval Christendom, and so without entering a new social "world" that cannot coexist with that "world."

In summary, I have identified three dimensions of systematic incompatibility between A and G. First, their respective theories are developed under incompatible strategies. Second, within a language, sufficiently rich to articulate the realm of daily life and experience, which functions as a metalanguage of both theories, a contradiction can be drawn between their theories. Third, social "worlds" in which the theories are respectively systematically applicable are incompatible.

There is also, accompanying the theory transition, not only a change of scientific 'world,' but also a change of social "world," from one where A easily applies to one where G applies well within the ascendent practices. This is well known. What are the connections between the theory and the social changes? Kuhn would acknowledge causal ones, for example: that the emerging social "world" provided material and social conditions needed for G to develop. Or, in the case of some later profound theory transitions, for example the chemical revolution of the nineteenth century (Hacking 1993), the new social "world" might be partly brought about by technological applications of the new theory. But Kuhn does not think that the connection has anything to do with explicating the rationality of the theory transition or of the change of strategy adopted. For Kuhn the trajectory of the tradition of scientific inquiry is essentially autonomous. While, for example, the conditions, sources of support, institutionalization and acceptance among the general public of the new strategy clearly need social explanation, the rationality of transitions from one scientific 'world' to another can be assessed principally in cognitive or epistemic terms, in terms of factors "internal" to the tradition – without taking into account the place of the scientific 'worlds' as "sub-worlds" of larger social "worlds," the applicability of theories in them, and historical transitions between them. I am not so sure.

FRUITFULNESS AND ADOPTING A STRATEGY

According to Kuhn, one rationally adopts a strategy because it is more *fruitful* than its competitors. A strategy's fruitfulness is assessed by its enabling theories to be developed that manifest the cognitive values to a high degree in the light of data selected (and generated) in its data gaining practices. After a certain time (by the time Newton made his contributions?), the greater fruitfulness of G's strategy became widely uncontroversial. This is compatible with incommensurability, for fruitfulness is judged within the framework of a given strategy. Within G's framework, one judges that theories are being generated that are manifesting the cognitive values increasingly to a higher degree; within A's, one judges that, for example, anomalies are multiplying. So, on Kuhn's view, no non-cognitive values are deployed in these assessments of comparative fruitfulness.

A, by his own lights, did not have to take into account all the new data accumulated by G, but he did have to address many of the criticisms made by Galileo in *Dialogue Concerning the Two Chief World Systems* in which Galileo appealed to phenomena (for example, the falling of a lighter and heavier body joined by a string) that were in A's purview, but which A could only deal with by use of auxiliary hypotheses that lowered the degree of manifestation of most of the cognitive values. That A faced anomalies (dealing with which lowered the degree of manifestation of the cognitive values) was well known. Galileo's criticism showed that they could be multiplied vastly. This is criticism conducted within A's own terms of discussion. That G also applied better to key objects of increasing importance in the realm of daily life and experience, as such, did not challenge that A applied well to the objects dealt with under its strategy, but it did put pressure on A by contributing to a growing sense of the insignificance of A. G also applied better to certain phenomena (for example, the swing, projectile motion, the movements of the planets) – under descriptions selected by those working under A's strategies, sometimes supplemented by categories deployed in common social practices – by incorporating them into a theory which manifested the cognitive values more highly of a domain that contains them. It is because A and G competed as theories of certain shared domains of phenomena that the increasing fruitfulness of G's strategies was inevitably accompanied by decreases in that of A's.

Aristotelian and Galilean accounts of planetary motions

I have emphasized that theories are accepted, not in an unqualified way, but with respect to specified domains of phenomena; and (Chapter 3) I maintained that if two theories conflict, choice between them can sometimes be made by appealing to the "comparative comprehensive" standard. If one of them manifests the cognitive values well with respect to a domain that includes that of the other, then *ceteris paribus* it is more acceptable, even more so if, from its perspective, the success of the other with respect to the smaller domain can be explained. Let D_A and D_G be the domains of phenomena of which A and G respectively manifest the cognitive values to a reasonably high degree. They only overlap; neither is contained in the other (so "comparative comprehensiveness" does not help to settle the dispute between A and G). Let D be the domain of overlap. The movements of the planets, moon and sun belong to D.[11] Furthermore, data concerning these phenomena, observational reports of the varying angular positions, brightnesses and other appearances of these bodies as obtained with naked-eye vision are selected to fit theories developed under both strategies. Attending to significant areas of incommensurability between A and G should not obscure that these theories are expected to manifest the cognitive values highly with respect to *some* of the same data. Both A and G wanted to "save the phenomena" of planetary motion, and there is considerable plausibility in those accounts that speak of the "empirical equivalence" (or equal empirical adequacy) of Ptolemaic and Copernican astronomy. How does G manifest the cognitive values more highly of D than does A?

It is well known that in A's tradition, Eudoxan was replaced by Ptolemaic astronomy largely because of the greater empirical adequacy of the latter (in relation to data describing the planetary phenomena of D) – even at the price of inconsistency with the fundamental posits of A's physical theory and thus significant reduction in explanatory power and internal (organic) unity, which is reflected in the common recourse to instrumental-ist interpretations of Ptolemaic theory. Given that A's physical theory is fundamental, that Ptolemy's theory has the degree of empirical adequacy it has is a mystery. How can posits strictly inconsistent with the fundamental posits of the cosmos predict better than posits (Eudoxan) that are consistent with the fundamental ones? And there are puzzling details: Why do the Ptolemaic constructions of each of the planets have components that are clearly linked with the sun's annual motion? And much appears to be *ad hoc*: the order in which the planets are arranged from the earth as center, which combination of geometrical devices (deferent, epicycle, epicycle on epicycle, eccentric, equant) works best in each case.

MacIntyre points out that from the perspective of the Copernican theory (certainly in its Newtonian version) all the mystery is resolved and the puzzling details are explained (1977). With the resources of the new theory (interpreted realistically), we can construct a narrative that explains how Ptolemaic theory could be empirically adequate within certain limits of precision, how those limits of precision could be improved upon, and why the theory needed the features it had. But from the perspective of the Ptolemaic theory we cannot construct a converse narrative. MacIntyre has convincingly argued that "source of interpretive power" (Chapter 3), being able to explain the strengths and weaknesses of a rival theory in a narrative is a highly ranked cognitive value, one that would be among the cognitive values endorsed by A (as well as G). Where it is highly manifested in a new theory (together with other key cognitive values), not only is it evident that the new strategy is more fruitful than the old (for dealing with certain domains of phenomena), but it is also explained why it is more fruitful. Therefore, a reason is provided to hold that the old will not be able to be rejuvenated in a way that might regain ascendancy over the new.

Consistent with the extent of incommensurability that obtains, and not applying standards for estimating the degree of manifestation of the cognitive values that A would not normally apply, we can affirm in summary:

1   G's strategies became more fruitful than A.
2   G's theory came to provide a better account of certain phenomena (and to fit shared data about them) that are of considerable impor-tance to A.
3   G's theory came to apply better to certain objects of increasing importance in the realm of daily life and experience.

For Kuhn, (1) and (2) provide sufficient *reason* to accept G's theory and to reject A's, and to discard A's strategies. I agree, though it is not a reason to accept that G's strategies have the power to generate theories that will encompass the phenomena, excluding those in D, to which A applied successfully. The argument appeals to the "comparative local strength" (Chapter 3) standard: since G manifests the cognitive values more highly of D than does A, *ceteris paribus* A is not acceptable of D_A.[12] That does not mean that A has been falsified. Through suitable adjustments of auxiliary hypotheses, A can be maintained consistent with its selected data; though, since these adjustments tend to be *ad hoc*, they generally bring about much lower manifestations of such highly ranked cognitive values as explanatory power. This logically permits the possibility that A may continue to adopt his strategy, and to accept theories developed under it of D, justifying his choice on grounds (social values, metaphysical commitments, religious faith?) to which fruitfulness – and ultimately any form of empirical inquiry – is considered subordinate. This is a logical possibility to which A, the Aristotelian scientist, can take resort only at the price of changing fundamentally the character of Aristotelian science as an activity, rooted in experience, seeking to discover and consolidate posits that manifest highly the cognitive values according to the most rigorous available standards of their estimation. It always remains possible to subordinate fruitfulness to non-cognitive values, but *within the practice of science* fruitfulness has primacy; to subordinate it to other values is to opt out of the "game" of science.

Fruitfulness: a necessary or a sufficient condition for adopting a strategy?

Kuhn maintains also that not only do (1) and (2) justify rationally discarding A's strategy, but also they justify adopting G's strategy rather than searching around for other strategies (which, if developed, might compete with G) – until such time as G itself may enter into crisis. For the sake of getting on with the "game" of science this makes a good deal of pragmatic sense: work in a 'world' where there is a community engaged in shared practices, where one can virtually be assured of progress rather than flounder around speculating in the darkness.

Attending only to pragmatic considerations, however, may obscure other pertinent matters, and uncritically presuppose answers to questions such as the following: Is G's strategy more fruitful than any other strategy that might have developed as a successor to A's? Was it (and only it) developed principally because of the historical exigencies of investigation? Did it (and only it) gain the opportunity to develop because disproportionate resources were devoted to its development, perhaps because of its potential interest of the bearers of certain social values? Were potential alternatives actively suppressed? Must there be a unique successor to A's, rather than a variety of competing strategies, each displaying fruitfulness – but developing theories which respectively become acceptable of different domains of phenomena that may only overlap one another? Might it be that the range in which G's strategies can be fruitful is essentially bounded (even if effectively unlimited) such that they produce theories acceptable of domains of phenomena (including those of D on p. 168) within the bounds; but that phenomena outside the bounds require for their investigation different strategies, but ones which grant an essential role for G's strategies in dealing with phenomena inside the bounds? Is it enough to respond to fruitfulness alone, and not also to take into account what are the domains of phenomena that can in principle become encompassed within G's strategies?

These questions are all obviously connected if not equally penetrating, but they may appear to be odd and strangely speculative. Instinctively one may want to respond: if there are alternatives to G's strategies, produce them and show what can be done with them; if there are bounds to their fruitfulness, define them and produce the arguments; otherwise, get on with the work of science. That would be the end of it if, in historical reality, the "game" of science were conducted in independence of its applications. But, in fact, science produces theories that are applied; and I doubt that the value of scientific activity can be articulated coherently without taking this into account. If I am right that the applications of G are particularly attuned to serve particular values, then the questions point to matters not only of social but perhaps also of cognitive significance.

The questions may be deemed inappropriate for other reasons too. One derives from the grip that materialist metaphysics has had on the self-interpretations of modern science, for it implies that material objects simply *are* such that they can be completely grasped with the categories of G's strategies. I have rejected this view in Chapter 6, where I suggest that the grip derives from the relationship of this metaphysics with the modern values of control. A second reason is that G's strategies have been extraordinarily fruitful; and, while particular fundamental theories that have enabled the general strategies to be interpreted concretely have been surpassed, the fruitfulness of the general strategies is confirmed by repeated development of fundamental theories that build upon the successes of their predecessors and resolve their anomalies. Fundamental theories have become remarkably wide-ranging, ranging across both experimental and "natural" phenomena and also the operation of the instruments which became essential in G's data gaining practices. That their range included the planetary motions has been of great importance. It demonstrated that a form of understanding could deploy categories derived from experimental practice *and* yet properly serve explanatory ends concerning natural phenomena.[13] This kind of fruitfulness, repeatedly confirmed, may appear to suggest that any talk of bounds or alternatives is merely skeptical. But G's strategies abstract objects from their human, social and ecological dimensions, and so its explanations do not instantiate full understanding. If there are bounds and alternatives, they might arise from these dimensions of objects. That the strategies are fruitful in the domains of astronomical and cosmological phenomena does not counter this, for these phenomena have no relevant human, social and ecological dimensions.

The underdetermination of later by earlier strategies

My questions are not so odd after all. They arise in the context of Kuhn's opinion that conflict among strategies occurs (and should occur) only after a dominant strategy comes into crisis, as reflected in its decreasing fruitfulness and the multiplication of its anomalies; and that a new strategy is (and should be) accorded hegemony in the relevant scientific community when it demonstrates its power to incorporate anomalies of theories developed under the old strategy into the domain of which its theories manifest the cognitive values highly.

Part of this opinion, with appropriate qualifications, can be readily defended. If, under the new strategy, the previously anomalous phenomena can be represented in a theory which manifests the cognitive values highly, there would be no serious point in seeking an alternative strategy for the sake of investigating the anomalies. This supports that the new strategy is a viable candidate to frame further investigation, but it supports only a hegemony limited to the domain containing the anomalies, and perhaps conservative extensions of it. Does the historical record support the factual components of Kuhn's opinion about the key role of anomalies in instigating and consolidating change of strategies, rather than matters connected with application doing so? Does it support that declining fruitfulness occasioned by the multiplication of anomalies, rather than insignificance of theories consolidated under the prevailing strategy, is the key to change of strategies? (A strategy may be fruitful, but the theories consolidated under it insignificant.)

In the present case the record is ambiguous because, while A's strategy was in crisis, from the perspective of ascendent values (including the modern values of control), its theories were also insignificant; and the development of G's strategies led, in one seamless movement reflective of their mutually reinforcing interaction with the modern values of control, both to the resolution of some of A's anomalies and to applications to objects that played central roles in the projects associated with these values.

As indicated in the previous paragraph, resolution of the anomalies does not support granting hegemony to G's strategy. Claiming its significance could do so. Conversely, affirming its insignificance (more likely its highly circumscribed significance) provides a ground to seek out other strategies, though not one to deny its fruitfulness or to claim that it is in crisis. Having such a ground to seek out other strategies is, of course, no guarantee that they will be formed; fruitfulness (in the middle to long run) remains a desideratum of the strategies that are adopted.

My point here is not that the history of science might have been otherwise. That is a truism. Rather it is that there is nothing inherent to the quest to gain theories that manifest the cognitive values highly that ensures a unique successor to a worked out strategy, or that prohibits that – paralleling the thesis of the underdetermination of theories by data – there is underdetermination of later by earlier strategies. A new strategy could gain a foothold by achieving success in attending to anomalies of the old strategy, or in attending to significant phenomena that do not fall within the purview of the old. Furthermore, even if one focuses on the anomalies, the strategies deployed directly in resolving them may not have the power to provide full understanding of the phenomena to which the theories, developed under the new strategy, apply. In the present case, G encapsulates well the material possibilities of the phenomena to which it applies, but one may query how well it encapsulates their possibilities when we do not abstract them from their human, social and ecological dimensions and, in particular, how well it charts the unintended side-effects of expanding and restructuring the place of these phenomena in the prevailing social "world." The centrality of resolving the anomalies of the previous strategies does not by itself explain that strategies, which abstract in this way, are considered adequate to guide research whose theoretical products are applied the realm of daily life and experience; the significance of the theories – deriving from the social value attributed to grasping more and more of the material possibilities of things – is also relevant to this explanation.

I do not think that there are resources within Kuhn's framework to counter my conclusion that significance as well as fruitfulness is needed to explain and justify the adoption of a unique strategy. It takes time for a new strategy to develop and demonstrate its fruitfulness. During this time ("revolutionary science"), a variety of competing approaches may be entertained. Then, according to Kuhn's account of the rationality of science, virtually "anything goes" as a new strategy is being sought; he provides plenty of historical documentation that, during revolutionary periods, many approaches are tried. Even so, it remains that G would not have developed if the necessary material and social conditions had not been made available for its development, and they may have been made available because of G's potential significance for the practices favored by the ascendent social values connected with control that I have described. On the same grounds, a potential rival strategy – either one applicable to a different class of phenomena deemed significant in the light of values in conflict with the insubordination of control, or one that incorporated G's solution of A's anomalies but limited the role of G's strategies to spaces where abstraction from human, social and ecological dimensions was judged appropriate – might not have received the necessary conditions to develop so that its fruitfulness could be displayed. Put another way, which new strategy (or strategies) comes to the fore may be essentially linked with the values of the social "world" which provides the material and social conditions for its deployment.[14] Demonstrating that a strategy is fruitful, and in a particular socio-historical context uniquely so, does not imply that it has shed the residue of the values that may have nurtured its development prior to the demonstration of its fruitfulness.

Since Kuhn supposes that a strategy has a unique successor, he does not explicitly consider this possibility. On his supposition there can be only contingent, causal links between the development of a strategy and the prevailing of a particular social "world." Some social "worlds," but not others, may provide the necessary conditions; but, once an old strategy has lapsed into crisis, which new strategy will emerge (if one does emerge) is not a function of prevailing social "world" and the values it embodies. For Kuhn, the trajectory of the scientific tradition remains *autonomous*; values do not play a role in the making of consolidated judgments of theory choice or strategy adoption. Fitting this picture, social values play no role in supporting the judgment that G is fruitful, and A not – so that if that is the key judgment at times of strategy change, values are not among the grounds for strategy change.

This picture, however, leaves as a merely contingent matter the especially salient applicability of G within projects in which the modern values of control are deeply embodied. Thus, while it expresses *autonomy* explicitly, it implies implicitly the lack of *neutrality*, ultimately an internally unresolvable tension. My alternative picture relieves the tension, since it recognizes that *impartiality* (though not *autonomy*) can coexist with lack of *neutrality*. Theories are evaluated in accord with *impartiality*, while produced under strategies adopted because of their significance with respect to particular values, as well as because of their (potential) fruitfulness. Given the possibility I have described on p. 174, values (and/or metaphysical commitments) would be an essential factor in strategy change. Only appeal to them would ground limiting systematic empirical inquiry exclusively to that conducted under an adopted strategy, in the case under discussion limiting the focus of inquiry to the material possibilities of things rather than also addressing their possibilities when we do not abstract from their human, social and ecological dimensions.[15]

This might suggest that the scientific community, if motivated by the value of *neutrality*, ought to entertain the simultaneous development of conflicting strategies so as to avoid that scientific research at a given time become essentially linked with particular social values. Conflicting strategies would be competitors, however, not only for adherents but also for material and social resources. In order to be deployed fruitfully, any strategy must gain its appropriate material and social conditions from a social "world," generally one in which the theories developed under the strategy are applicable. But the conflicting strategies might be dependent upon "worlds" which exclude each other, as do the "worlds" of medieval Christendom and modernity, and the "worlds" of contemporary globalization and the free market and Latin American popular organizations (Chapter 8). Then, Kuhn might be right that coexisting competing strategies are impossible; not because the objective of science (to produce theories that manifest the cognitive values highly) requires that it be so, but because the social "worlds" upon which implementing the strategies are dependent mutually exclude each other. This would confirm that the fruitfulness of G's strategy is only a necessary condition for its rational adoption in the scientific community; and that power (and related social values) are necessary for its exclusive adoption, so that the exclusive adoption of a strategy reflects not only historicity, but also power – most effectively power whose associated values are articulated in a dominant consciousness that lacks the conceptual resources to render an alternative intelligible.

Fruitfulness and applicability in the realm of daily life and experience

Let us return to the applicability of A and G in the realm of daily life and experience. This realm changes with history, partly under human causal agency as human beings often radically modify the material and social conditions they inherit, so much so that people in different epochs think of themselves as living in different "ages" ("worlds"). With the movements of history, there come into being different realms of daily life and experience whose objects, phenomena and significant modes of interaction are considerably (though far from completely) different. Which theories are applicable may depend on the realm of daily life in question; and changes of realm may engender changes of applicable theory. Moreover, the development of theories and the practices of applying them – activities in the scientists' 'worlds' – require conditions from the realm of daily life. The scientist's 'world' is always inserted in a realm of daily life and experience, in a social "world."

G's 'world' could not coexist (except as a minor fragment) with the "world" in which A applied easily, if only because that "world" did not make available the technological conditions for the increasingly necessary instruments to be manufactured. The instruments, products of the new technology, were not just external conditions for the development of G's science (like funding and political support). They became an integral part of the object of experimental investigation (see p. 159–60; also Chapter 6). G intervened with their aid, and observed the interventions and their outcomes. So, if G's supporters in practical life had acted on the maxim, "do not pursue the implementation of G's possibilities further until the issue between G and A has been settled decisively by scientific inquiry," the empirical data needed for the eventual, decisive support of G would not have been obtained. Neither G's scientific project nor the technological project could develop significantly in a "world" in whose realm of daily life A's theory was considered to apply satisfactorily to its most important practices and phenomena. In such a "world," key objects needed for G's experimental program (that is, objects that belong integrally to its domain of inquiry) do not come to the fore, and the realization of G's possibilities (except incidentally and in individual instances) is generally considered undesirable. At the time of the transition between A and G, the "world" of medieval Christendom was breaking apart, and increasingly the "world" was coming to contain objects of growing importance in the structuring of daily life of which G, but not A, could provide some understanding (for example, canons, mechanical clocks, the mechanisms of the printing press, pumps, optical and meteorological instruments). Nevertheless, those who deemed G's possibilities generally desirable could not yet appeal to the decisive empirical support of G to ground the genuine possibility of further developments and implementations of these possibilities. During the period of the transition it was important, for the eventual consolidation of G, that the technological project develop, driven both by its values and by the inductively-derived confidence that the current developments did not represent the limit of development, even if the limits of the possible were not very clear and A implied that they were highly circumscribed – how the practice turned out could settle that! G, I suggest, was *inconceivable*, lacking internal conditions and not just external material and social conditions, apart from the developing technological project. And the more the technological project rose successfully to prominence in a society, the less would be the space remaining for the application of A in the realm of daily life, so much so that A has practically no applicability in the realm of daily life today.

On the other hand, even retrospectively and speculatively G would have little applicability in a "world" where A applies easily. (It is not just that G could not develop in such a "world.") First, concerning the characteristic phenomena of the realm of daily life of such a "world," those highlighted in the modes of life where the dominant articulations of the "world" were widely shared, it is doubtful that one could plausibly affirm that, as a whole as distinct from incidental particulars, G explains them better than does A – especially when we remember that here the control of nature was considered subordinate to such relations as attunement, that *the world* (cosmos) was conceived as teleological through and through, and that the characteristic phenomena were considered concretely in their multifacet-edness not abstracting from some of their aspects. Second, with limited technology and not much interest in it in the dominant modes of life, this realm lacked the objects and practices where the strength of G in informing activity is located. Retrospectively, G could be applied to the astronomical phenomena which A aimed to grasp. This seems to be the one clear exception to the generalization stated at the beginning of this paragraph. Moreover, G manifests greater empirical adequacy with respect to these phenomena. But it is difficult to see how that greater empirical adequacy could have been demonstrated without developments of G's research program, including its deployment of optical instruments, and certainly the demonstration that G manifests the other cognitive values highly with respect to these phenomena requires the embedding of G's account of the planetary motions in a broader theory that manifests the cognitive values highly with respect to experimental and mechanical phenomena that would not figure prominently in a "world" where A applies easily.

With respect to the totality of phenomena to which A applied easily, A manifests the cognitive values more highly than did G, so that in a "world" where A applied easily, G could not supplant A. But G's manifesting the cognitive values to a higher degree of an important sub-domain (the planetary motions) of this totality does undermine the reasonableness of accepting A as a theory of the totality. With the changing social "world," many of the phenomena of this totality ceased to be significant in the realm of daily life and experience. So there was little reason left to attempt to shape a strategy which would aim to generate successor theories to A that might manifest the cognitive values to a sufficiently high degree of the old totality (in the light of the higher standards of appraisal brought about by the competitive situation that came into being with the introduction of G). It does not follow that G's strategy – although it is demonstrably fruitful

– can encompass all the phenomena in the emerging "world" that may be of interest in the light of sustainable social values that conflict with the dominant or ascendent ones.

I have suggested that the factor of applicability, in addition to fruitfulness, is a key consideration in the widespread adoption of a strategy. And applicability is essentially linked with social values. Initially, G's strategy gained the social and material conditions to develop because of its promise linked with the modern values of control, and its capability to provide understanding of phenomena of special interest to the ascendent groups, which began to embody these values. (Because of that link, and thus the threat it posed to the prevailing social/religious/political/economic structures, it also had to face efforts by the then dominant powers to suppress its development.) Eventually the social "world" became so dominated by the phenomena that G could explain well that A lost its role in practical life. G's strategy is "rationally" applied in the "world"; to a large degree, it is not possible to live (function efficaciously) in this "world" unless one's practice is informed by G – though, consistent with this, its deployment may be limited within identifiable bounds and granted a subordinate role in relation to other strategies.

Which "world" is the swing in? Depending on its socio-historical place, it could be in any one. It is also in *the world*. But, given the general character of human experience, we have no way to grasp objects of *the world*, except insofar as they are also in our "world." In our "world," objects are objects of value, objects understood in relation to our experience and practice, valued more or less and in different ways depending on their place in the realm of daily life and experience. We adopt strategies, subject to fruitfulness as a necessary condition, (in part) in virtue of their capability to develop theories that provide understanding of those objects that we consider to be exemplary objects of value. Kuhn's account disguises this by mistaking a necessary condition for adopting a strategy (one articulated fully in terms of the cognitive values) for a sufficient one.

This argument reinforces my view of the mutually reinforcing interaction between the materialist strategies (G) and the modern values of control. It explains the virtually exclusive adoption of this strategy, while opening up the possibility that there are rival, fruitful strategies to be found dialectically linked with opposing values. It also provides richer content to Kuhn's view that A and G observe in different 'worlds,' different social "worlds" that structure the realm of daily life and experience, as well as scientific 'worlds' grasped under different strategies. We turn now, in the next two chapters, to explore concretely the prospects of two anticipatory alternative strategies.

8 A "grassroots empowerment" approach

That prioritizing the materialist strategies needs explanation has been at the core of the discussion of the previous two chapters. I have argued that the relevant explanation cannot be derived simply from appeal to the general objective of gaining understanding of phenomena. Instead, it follows from mutually reinforcing interactions that exist between inquiry conducted under the materialist strategies and the modern values of control. In the course of the argument, the suggestion has repeatedly surfaced that there may be other strategies, alternatives to the materialist strategies, with mutually reinforcing interactions with value complexes that clash with the modern values of control, that under the appropriate material and social conditions might frame fruitful scientific (systematic empirical) inquiry. Clearly the argument would be strengthened if some alternative strategies were to be concretely identified and if, at least in an anticipatory way, their potential to produce theories that manifest the cognitive values to a high degree were displayed.

Not only would this strengthen the argument it would also have considerable social importance. The products consolidated under the materialist strategies are not *neutral*, being especially significant for value complexes containing the modern values of control. Pursuing inquiry under alternative strategies might generate knowledge of significance for rival value complexes and, in doing so, open up the possibility for the aspiration to be regained that the general practices of science, considered as including a place for a variety of strategies, might come to manifest *neutrality* in a more robust way (Chapter 10).

The general objective of science, to gain understanding of phenomena (O), does not lead immediately or (in principle) uniquely to adopting the materialist strategies (Chapter 5). Under these strategies there are encapsulated possibilities of phenomena that can be identified in terms of the generative power of underlying structure, process and law and thus, largely equivalently (Chapter 6), in terms of their potential value for value complexes that contain the modern values of control. But there is no a priori or empirically well supported reason to consider these possibilities to exhaust the possibilities of phenomena. Thus, I proposed considering the materialist strategies – the strategies under which there are encapsulated possibilities of special significance where the modern values of control are adopted – to define a particular approach to scientific inquiry (the Galilean/Baconian, O_1/O_1', approach) which I rephrase as follows:

O_1 The objective of the Galilean approach to science is to represent (in rationally acceptable theories) the order (structure, process and law) that is posited to underlie phenomena, to represent phenomena and possibilities in terms of being generated from the posited underlying order, and thence to discover novel phenomena.

O_1' The objective of the Baconian approach to science is to encapsulate (reliably, in rationally acceptable theories) the possibilities of domains of phenomena that are of potential value for projects that express the modern values of control, and thence to discover means to realize some of the hitherto unrealized possibilities.

Recognizing that the Galilean and Baconian approaches together constitute a unity is helpful. It precludes objecting to a proposed alternative strategy simply on the ground that adopting it is motivated by holding a particular set of values, and it opens up that O_1/O_1' may be considered as an instance of a whole cluster of alternative approaches (O_i/O_i'), where each O_i' would be an instance of the schema:

O_i' The objective of the (…) approach to science is to encapsulate (reliably, in rationally acceptable theories) the possibilities of domains of phenomena that are of potential value for projects responsive to the value complex (…), and thence to discover means to realize some of the hitherto unrealized possibilities.

Is introducing the O_i' more than a formal exercise? Are there really alternatives to O_1/O_1'? Can strategies be identified (1) which are not reducible to the materialist strategies (though they may complement or subordinate them), (2) which interact in mutually reinforcing ways with value complexes that conflict with the modern values of control, and (3) under which theories can be developed and come to be accepted in virtue of their manifesting the cognitive values to a high degree? An alternative strategy must have all three of these characteristics, and the second does not guarantee the third. I have suggested (end of Chapter 6) that one might look for alternatives at the margins of the advanced industrial societies and among minority movements in mainstream institutions. In this chapter, I will follow up this suggestion with detailed consideration of an approach that I call a "grassroots empowerment" approach (O_2/O_2'); and, in the next chapter, I will explore a feminist approach (O_3/O_3'). The upshot will be vindication that there are alternative strategies, albeit with vastly different degrees of development and numerous ambiguities about their promise for further development. Nevertheless, they possess sufficient cognitive credentials to infuse a certain moral urgency and complication into the question: Which strategies to adopt? Which approach to follow?

CONTESTING THE MODERN VALUES OF CONTROL

Let us return to the question of human flourishing, which remains – in my opinion – the touchstone of all practice and inquiry. In the modern viewpoint, the modern values of control are said to enhance human flourishing. But that is contested, for example, by many feminist and environmental perspectives, and especially by numerous popular, grassroots perspectives throughout the impoverished sectors of the world.

I will work towards the definition of a "grassroots empowerment" approach to engaging in scientific (systematic empirical) inquiry in two steps. The first is to discuss different notions of "development," for the contestants challenge the social values of the leading institutions of "development," seeing them more as among the causal contributors than as remedies to the vast suffering and misery experienced by the impover-ished. This is the background to their rejection of the modern values of control. They do not agree that expanding our capability to control nature is able to address properly the reality they face; or that its associated forms of understanding in general (as distinct from when subordinated within an appropriate form of full understanding) enable the identification of possibilities that might contribute to a social transformation that would serve their communal ideals of human flourishing, and further the manifestation of such values as cooperation, participation, assuming responsibility for the future, solidarity, self-reliance, respect for nature and the dialectical unity of means and ends. The projects of the contestants need to be informed by understanding – forms of systematic empirical (scientific) understanding. I will then, the second step, elaborate this point by considering various ways of conducting research on "the seed." All the issues under discussion come to a focus in it. An additional aim of the discussion of the seed is to illustrate further how research conducted under the materialist strategies does not produce *neutral* theoretical products, thus laying the groundwork for revising the formulation of *neutrality* (Chapter 10).

Concepts of "development"

"Development," like its predecessor "progress" and like "freedom" and "democracy," is a key term in the contemporary lexicon of legitimation. These terms are deployed in order to legitimize social goals, and to affirm the rationality, the practicality and the realism of their adherents' practices and policies (Lacey 1991a). The meaning of all of them, however, is contested and so they can be sources of ambiguity, misunderstanding, futile controversy, and paradoxically rationalizations for maintaining and deepening conditions of ecological and social devastation. Fundamental terms of the lexicon of legitimation are contested because their logical, theoretical and methodological force derives from the complex interaction of three factors:

1   moral ideals;
2   concrete embodiments of the ideals, for instance actual strategies, processes, institutions and policies which embody the ideals only more or less; and
3   theoretical idealizations.

The contestation can sometimes be obscured, even suppressed, because of the power associated with a hegemonic interpretation.

In its hegemonic interpretation, "development" represents such moral ideals as individual freedom and the overcoming of poverty, as well as the further manifestation of the modern values of control. It is regarded as embodied to an acceptable degree in the advanced industrial societies, which, in turn, are typically characterized under the theoretical idealizations of democratic capitalism. Clearly, at the present historical moment, the forces that bear this interpretation have gained unprecedented power.

Nevertheless, the contestants remain (Fabián 1991), some even challenging that there is an acceptable ideal of development (Escobar 1995). They challenge the predominantly individualistic ideal on grounds such as that the embodiments offered by the advanced industrial societies cannot be universalized, that it represents the value of autonomy without adequate balance with that of solidarity, and that it abstracts individual from cultural identity. They also challenge the theoretical idealizations maintaining that they do not provide adequate explanatory models of current economic and social realities, proposing instead various versions of dependent capitalism (Lacey 1985). Thus they identify certain capitalist orientated development projects in the impoverished countries not as means to overcome poverty, but as causes of underdevelopment. Contestation of "development" is probably inevitable at the present time. On the one hand, the aspiration to share the way of life characteristic of the advanced industrial societies has obvious appeal; but, on the other hand, its institutions are (and probably will remain) implicated in the deep and multi-dimensional sufferings of perhaps the majority of people in the impoverished countries. "Development" is the site of on-going tension between aspiration and what is possible, a tension that can be relieved in the light of the outcomes of empirical research and the achievements of concrete practices and movements, but one that ensures that there cannot be a sharp separation of fact and value at this site.

The issue of development demands attention because very large numbers of people in the Third World and elsewhere experience the condition of their lives in all of its dimensions – material, social, agentive, spiritual, psychological, cultural – as diminished, distorted or even intolerable; and they have come to realize that it need not remain that way. Development represents both a negation of their present condition, and a process of transformation. The contestation of "development" concerns the appropriate goal of the process of transformation, that is to say, the appropriate specific form that the negation of their present condition should take. How is this goal chosen? Who chooses it and who ought to? Once the goal is chosen, what is the appropriate process of transformation? What is the relationship between process and goal?

A distinction between "modernizing development" and "authentic development" is at the core of the contestation of "development."[1] These two ideal types differ in how they represent the negation of the present condition of impoverished peoples that expresses the goal of development, and in how they identify the processes of transformation. The negation may be variously understood depending on whether the state of development or the current condition of the impoverished is taken to be well defined.

For *modernizing development* it is the state of development that is considered well defined: it is represented by the institutions and values hegemonic in the advanced industrial countries, and the processes of development involve economic growth, industrialization, modern technology transfer, integration into the world capitalist economy, etc. The current condition of impoverished peoples, then, is characterized as "under-developed." Development is the negation of underdevelopment (the negation of the negation of development). "Modernizing development" is widely understood and needs no further elaboration here. "Authentic development" is not widely understood, so I will elaborate it in some detail.

Authentic development

For *authentic development*, what is taken to be well defined is not the state of development, but the current condition of the impoverished, which can be empirically charted and theorized in terms of such notions as oppression and dependency. For it, development gradually gains definition by negation, through political and social action and organization, of the various dimensions of suffering experienced by the poor. Authentic development is meant to be a response to the concrete and multidimensional sufferings of large numbers of people, especially in the impoverished countries, and to provide a means towards their negation. These sufferings obviously involve a material (bodily) dimension whose intensity often overwhelms one's consciousness of the other dimensions. They also have a social dimension as people experience the disruption of their families and communities, the necessities of migration, and a sense of isolation – to which may be added special racial and gender components. On the cultural dimension, there are the sufferings derived from perceiving one's traditions, cultures, languages, histories and ecologies being destroyed. Then there is a sense of powerlessness and helplessness, sometimes demoralization and depression, the threat of nihilism, and the recognition of the unfreedom of one's life: that one is subject to the pushes and pulls of forces outside of one's control and often understanding (because of the cultural destruction and the denial of education), that one's own perceptiveness, values and agency play no role in the unfolding of history. The sufferings may also involve the early and painful deaths of one's children, the woes of unemployment and unstable intermittent employment, daily confrontations with drugs and violence, the devastation of being driven from one's land, experiencing the contempt (and fear) of the powerful and well-off, and the experience of violence exercised against those who organize for change (Lacey 1991b).

On articulation, the negation of the multidimensional sufferings of the poor provides a conception of a full or flourishing life, and supports the aspiration for a social order in which free and flourishing lives can be led by as many as possible. It does not require a radical rupture with traditional cultures, and, in many cases, it draws from them conceptions of social justice that seek to embody such values as cooperation, widespread participation, self-reliance and respect for nature. It is at ease with cultural diversity and the expectation that from different cultures different positive definitions of development will emerge. Its measure, thus, cannot be material progress or economic growth *per se*, and relatively independent and unhindered technological innovation cannot be its driving force. Rather it seeks to integrate economic growth with the poor claiming their human agency and with the unleashing of their capabilities for exercising responsibility in shaping the conditions which structure their lives. The goal of authentic development includes its means: to generate a process of negating current impoverishment, which integrally includes material and economic dimensions in balance with others, informed by the understanding of the poor themselves and with their active participation. The possibilities, hoped to be realized in this process, clearly go beyond those encapsulated in current first-world institutions and subsumed under its regularities. They include fuller manifestations of values that fit uncomfortably with those linked with modernizing development (Lacey 1997a): solidarity in balance with individual autonomy, social goods ranked above private property and profits, the well-being of all persons above the market, the strengthening of a plurality of values in place of commodification, human liberation in balance with individual liberty and economic efficiency, the rights of the poor above the interests of the rich, taking responsibility for the future instead of resignation in face of the projects of the powerful, democracy enriched with participatory mechanisms and not limited to formal democracy, and the proper balance of civil/political and social/economic/cultural rights. In summary, the values of enhancing local well-being, agency and community – or of "grassroots empowerment." Reflecting this, their central practices are located in the social movements of the poor themselves.

Popular organizations

In Latin America, the relevant social movements are often called "popular [or grassroots] organizations"[2] and collectively they are referred to as "the popular movement." Popular organizations (*as ideal type*) are characterized, in the first place, by practices (forms of "struggle") that both derive from and reinforce the group identity of, for example, women, landless families, indigenous peoples, workers, or refugees. They are struggles which tend to focus on concrete objectives that address the self-identified needs and interests of members of the organization, concerning for example: self-reliance, education, health, housing, land tenure and basic rights (articles by Martín-Baró in Hassett and Lacey 1991). Second, they are grassroots organizations with grassroots leadership, styles, visions, values, language, knowledge and cultural forms; and at the same time they are open to coalition politics with elite organizations and their members, recognizing that any significant social change will have to involve a role for such organizations and will have to involve a refocusing of them.

Third, especially when they enter into broader (municipal, regional, national, international) deliberations, the popular organizations tend to articulate their objectives in the language of human rights.[3] Indeed, the Human Rights' Movement and institutions for the defense and propagation of human rights have become an important part of the popular movement (Lacey 1991b). They emphasize everyone's right, not just to life, but to *a* life worthy of a human being, opening for ongoing reflection, for example, the issue of the relative ranking of the rights to property and to meaningful work. They do not articulate their objectives in terms of a particular political or economic system, capitalist or socialist. Struggle, for them, is to bring about the fullest possible embodiment of human rights, not necessarily, for example, to usher in socialism. Bringing that about, of course, entails structural transformation. What the desired social structures will be that adequately embody human rights, however, is not defined theoretically in advance; but will emerge from the necessities and creativity of the struggle, reflecting the inputs and actions of those engaged in it.[4] Within the popular organizations, justice tends to be defined in terms of an adequate embodiment of the full range of human rights, and democracy is above all respect for human rights – a respect usually held most likely to be present within structures of representative government and separation of powers.

Finally, popular organizations represent an "organic" unity between means and ends, and between ameliorative action and structural transformation. They represent in anticipation the values that they desire to have embodied throughout society. Their movement towards new structures involves: a) growth whose claims and appeal are grounded in the partial realizations already actualized in the organizations (Lacey 1997c; and b) a keen sense of the dialectic of personal development and social change. The desired new structures cannot be defined or created from above or from outside (though they cannot be created without alliances with and support from other groups); nor can they be created by a violent cataclysmic event. The mistake of many revolutionary movements has been to hold that liberation comes from an armed struggle (which at most can remove obstacles to liberation) rather than from the practices of the popular organizations in all of their variety. The mistake of the modernizing elites is that liberation comes from the imposition of the structures of "developed" societies.

Development, science and technology

The place of science in modernizing development is clear enough. Such development requires the availability of modern (materialist) scientific knowledge, scientific institutions, and the dialectic of science and advanced technology. Moreover, materialist scientific understanding itself becomes a value for it, even where it transcends possible applications (Chapter 6). This clarity, however, sits side by side with an ambiguity when we address "developing" societies – whether, while developing, they need their own independent institutions for generating (as distinct from transmitting and applying) scientific knowledge. Here *neutrality* is often appealed to, suggesting that the place of origin of scientific knowledge is irrelevant to the assessment of its cognitive credentials and to its significance in application; then, in the name of efficiency and ready availability of resources, arguing for the concentration of basic scientific research in the advanced industrial societies or institutions supported by them. The consequence, then, is that science may become yet another instrument for entrenching dependency (Bunge 1980; Lacey 1994).

There are good methodological reasons to *distinguish* between fundamental (basic, pure) and applied scientific research. Nevertheless, since scientific research is conducted in institutions, pure and applied are never fully *separated*. When we emphasize fundamental research, we are pointing to the definitions of research problems and the realms of possibilities to be explored, which arise from the internal unfolding of research programs (that are actively being followed under the direction of particular strategies), aspects of scientific practice that are linked with the ideas of *impartiality* and *neutrality*. Here research is conducted in abstraction from immediate concern for application, but that does not imply the absence of real, mutually reinforcing interactions between that research and applied interests. This heightens the interest in the question of what alternative forms science (including fundamental science) might take in impoverished countries so that research could be conducted under strategies consistent with the ideals of authentic development rather than exclusively with those that maintain mutually reinforcing interactions with the modern values of control.

Scientific understanding gained under the materialist strategies, the exemplary instance of wide-ranging understanding, is especially apt for informing advanced technology, the core instrument of the modern values of control. A version of full understanding (Chapter 5) may also inform a type of technology, *appropriate technology*. By "appropriate technology," I mean any technology that serves the interests of authentic development.[5] Appropriate technologies consist of material objects (together with associated techniques and bodies of "know-how") that have been made by human beings for the sake of increasing those forms of human control over natural objects which will contribute to the greater well-being or flourishing (in all dimensions) of all human beings, but especially of the poor who constitute the majority in the impoverished countries. Where appropriate technologies are sought, the value of augmenting our capability to control natural objects is subordinated to values linked with widespread human flourishing. This implies that the users, in community, of an appropriate technology have control over its production and use, and over its material conditions (such as the raw materials needed to make and operate it, and the services needed to maintain it), so that its production and use are directed towards meeting the needs of the community. An appropriate technology, thus, interacts dialectically with relations of production that encourage universal participation.

Appropriate technology differs from the dominant technology in two related ways. First, it is characterized by social relations that dialectically further the well-being of the poor majority, rather than dialectically generate class inequalities and the tendency to privilege the interests of such groups as the rich and the military. Contrary to common viewpoints, the dominant technology is not neutral (Tiles and Oberdiek 1995), since the capabilities it generates to exercise power over natural objects empower some human beings at the expense of others. Both the dominant and appropriate technologies are linked directly and conceptually with social values and interests; they differ in the specific values and interests with which they are linked.

Second, appropriate technology is explicitly informed by versions of full understanding, often involving the interplay of "technical" and local knowledge, or the systematic development of local knowledge. It responds to questions like: "How can we produce food so that all the people in a given region will gain access to a well-balanced diet?" rather than to: "How can we maximize food production under 'optimal' material conditions?" Answering such questions requires a mode of investigation that, unlike materialist research, is not restricted to analyzing food production as a function largely of quantitative variables, and that does not consider separately the technical, the biological and the sociological, or production and distribution variables. The investigations that inform appropriate technology may and often will be informed by the results of materialist research, though interest in and pursuit of that research will remain subordinated to the objectives of authentic development, and it also will respond to specific questions about the links of crop yields and the variety of products gained from the crops with a range of socio-economic and locally specific variables (as detailed on pp. 194–7). Full understanding – insofar as efforts to gain it follow strategies that interact in mutually reinforcing ways with the values of authentic development – tends to produce not general theories, but local profiles, structures and narratives, with generalizations often sharply bounded in application to the local domain. Furthermore, its generation and consolidation may require participation from the local grassroots in interaction with the "expert" practitioner; but, as understanding, it is neither subordinate to materialist understanding, nor (in principle) inferior to it in status.

THE SEED

Earlier (Chapter 5), I considered the seed in two ways: as an object which generates, upon cultivation, crops with quantifiable yields; and as an object which is integrally part of social processes. Suppose we ask: How can we maximize the production of wheat under "optimal" material conditions? This question abstracts from the conditions of daily life and experience and the prevailing practical activities of the producers and consumers of the crop. It can be addressed as part of research under the materialist strategies (applying basic physical, chemical, genetic, biochemical and other scientific knowledge), in which crop yield is investigated as a function of such variables (which are open to quantification) as the use of fertilizers, insecticides, water, machinery and strains of seeds. Such research is typical of the "green revolution" and its biotechnology successors, whose practices are among the foremost bearers of the modern values of control.

The green revolution

The green revolution[6] is rooted in the phenomenon that low-yielding varieties of wheat (or other crop) may be grown in isolation, and from them may be produced hybrid seeds that generate (under the "right" conditions) high-yielding plants, whose yields are very much greater than those obtained from plants that are grown from regular, field-fertilized seeds. (The low-yielding, "pure" varieties are obtained by trial and error separation out from field-fertilized seeds, which have been developed over the years informed by traditional knowledge.) The phenomenon has been widely replicated, and its applications have been widely acclaimed.

Green revolution practices generally increase crop yields (and, often, national exports and corporate profits) at least in the short run. They also have significant costs especially as calculated from the perspective of those who give high ranking to the social values of social and ecological stability and to the enhancement of local well-being. In the first place, production has become vastly more capital intensive and it requires expensive inputs: hybrid seeds, water (irrigation), fertilizers, pesticides, herbicides, machinery and the energy to run it. Second, there have been side-effects of production with negative environmental and social impact. Environmentally, with variation from case to case, there has been depletion and poisoning of soils, loss of diversity in the gene stock of crop seeds, disruption of streams (and other negative effects of dam construction), desertification, increased dependence upon fertilizers, herbicides and pesticides, and reduced quantities of other outputs of traditional crop production. Socially, small-scale farming has declined, leading to migration to cities with accompanying unhygienic and psychologically threatening living conditions and increased homelessness, increased unemployment and underemployment, deepening of dependence on international capital (both for imports of fertilizers, technology transfer, etc., and for markets) – in short, social disruption and consolidation of market economies that tend not to cater to the needs of the poor majorities, and that have been especially destructive for women and children (Shiva 1989).

Regarding one of the most famous implementations of the green revolution, Shiva has summarized the costs as follows:

> Instead of abundance, Punjab has been left with diseased soils, pest-infested crops, water-logged deserts, and indebted and discontented farmers … conflict and violence. … [E]cological and ethnic fragmentation and breakdown are intimately connected and are an intrinsic part of a policy of planned destruction of diversity in nature and culture to create the uniformity demanded by central management systems.
>
> (Shiva 1991: 12, 24)

Shiva (1988) has generalized these remarks and affirmed that what she calls "reductionist science" (roughly: science conducted exclusively under the materialist strategies) contributes (together with other factors) to producing a four-fold violence.[7] First, there is violence against "the beneficiary of knowledge," for example, farmers in rural India. Second, against "the subject of knowledge": granting "monopoly" to knowledge gained under the materialist strategy devalues the knowledge of the bearers of other forms of understanding and the activities informed by them, and poses no barriers to social and economic projects which render these forms insignificant and thus diminish the agency of their bearers. Third, and relatedly, against "knowledge itself": when in the name of sound "scientific knowledge" traditional knowledge is not only devalued, but also suppressed and distorted.[8] And finally, there is violence against "the object of knowledge": when, for example, a project informed by reductionist knowledge destroys "the innate integrity of nature and therefore robs it of its regenerative capacity" (Shiva 1988: 232–5), or destroys the genetic heritage of a region.

Commodification of the seed

Perhaps the most striking change brought about from adopting in agricultural practice the knowledge gained from green revolution research, and one that exacerbates all the alleged costs listed on p. 191, is that the seed once the common heritage of humankind – a biological entity normally generated each year as part of the crop – has tended to become a commodity (Shiva 1991, 1993; Kloppenburg 1988). Indeed, the significance of the research products, and thus the provision of the conditions for carrying out the research, are largely tied to this tendency (Rouse 1987). Although the seed's tending to become a commodity is both a condition and a consequence of the practical application of green revolution knowledge, the possibility that it become a commodity is not encapsulated in theories generated under the materialist strategies, since becoming a commodity involves the social relations of the seed. This tendency is being furthered through the use of patents and appeal to intellectual property rights that have marked the new biotechnology revolution (Suárez 1990; Kloppenburg 1988; Brush and Stabinsky 1996; Shiva 1997). The scientific research that has informed the green revolution – conducted under strategies needed for investigating the relationship between the magnitude of crop yields and physical and chemical inputs – has led, when adopted in practice, *both* to increased yields in the short run *and* to the commodification of the seed.

Furthermore, with the commodification of the seed third-world agriculture has become more inserted into the international economy in ways that serve the special interests of agribusiness, a sector of land owners and some related industries, bearers of the modern values of control who are linked integrally with particular socio-economic values. Given the material and social conditions of the research, it could not have been otherwise; it has served the interests of the market rather than other (opposed) social values; it could not be made to serve the interests of all value complexes. Theories, soundly accepted within the strategies deployed in green revolution research, are not generally significant across viable value complexes. I take this to be a fact, not *per se* a point of criticism of these theories. Criticism depends upon adhering to a value complex for which they lack significance; and then the criticism is that they lack significance, not that they lack the proper cognitive credentials – that green revolution knowledge is not *neutral*.

Is green revolution knowledge neutral?

Those who defend *neutrality* might counter that, in the light of contemporary realities, the value complexes for which these theories lack significance are not viable (and not just presumed to be not socially sustainable: Chapter 4). Certainly the adherents of the modern values of control look to further technological (for example, biotechnological) developments to reverse or prevent further damage, and regard the consequent social reorganization as necessary for development; for them development (modernization) always requires deeper implementations of control. They may elaborate their objection by maintaining that I have ignored the obvious and urgent fact that the increases in crop yield were (and are) necessary to avert hunger in a world with a rapidly increasing population. Therefore, they would insist, the green revolution has contributed to meeting the most basic human need, a universal value, which must be a part of any viable value complex. This is not an implausible view. Put like this, however, the objection is simplistic, for it does not take into account the relationship between the mechanisms and institutions of production and those of distribution (and those of research). It has been claimed that, while the green revolution has provided conditions for many more people to be fed, it has also produced a redistribution of the hungry (Shiva 1991), and it has not produced social mechanisms that are adequately responsive to the basic human needs of large numbers of people (or respectful of their rights). But, however one puts the objection, it rests upon a crucial *presupposition*: there are no mechanisms to increase crop yields in the needed quantities outside of the green revolution technologies (and their biotechnology successors). In particular, it presupposes that traditional agricultural mechanisms cannot be improved through research so that they become capable of increasing yields significantly; it presupposes that they cannot approach even remotely the efficiency introduced by the green revolution. Put another way, the presupposition is that the green revolution has occasioned no relevant *lost possibilities*.[9]

Although "no lost possibilities" is apparently a "factual" presupposition, the policies and practices of the green revolution have rarely reflected any empirical investigation of it. Research conducted at the predominant scientific institutions on these matters reflects antecedently made policy decisions, rather than *vice versa*. That this is not widely recognized reflects how deeply this presupposition is ingrained in the thought prevailing in the advanced industrial countries. It is part of the a priori thinking that sees development (modernization), and thus the further manifestation and embodiment of the modern values of control, as the "solution" to all problems. But, one might object, is it not empirically supported by the clear fact of recurrent shortfalls in food production in the post Second World War years? Does not this fact speak definitely to the inadequacy of traditional agricultural technologies? Shiva suggests an alternative explanation of this fact in the commercialization of agriculture (a product of colonialism), and in the breaking of the tie between the traditional technologies and their social relations (1991: 26). The implications of this suggestion become clearer in the light of the distinction she uses between two kinds of poverty: "poverty as subsistence" and "misery as deprivation" (Shiva 1989). The deprivation, from which development is meant to be the solution (underdevelopment), Shiva maintains, is itself caused by the process of development. Moreover, serious investigators have argued that there is some *evidence* that indicates that the presupposition of "no lost possibilities" is false.[10] Some of them maintain that the hybrid seeds were not necessary to produce significantly higher yields (Shiva 1991; Levins and Lewontin 1985). According to the same underlying genetic theory that informed the hybrid seed research, they maintain, comparable yields could be obtained from appropriately selected "pure" (non-hybrid) varieties and used in ways more compatible with social and ecological stability, but contrary to the interests of modernizing development. Pending further investigation, it appears that the objection that the value complexes, from which criticism of the green revolution is marshalled, are not viable rests on a presupposition which has not been accepted in accordance with *impartiality*. That objection therefore fails.

"Lost possibilities": agroecology and the green revolution

Those, often connected with grassroots movements, who adhere to value complexes in which, for example, enhancing local well-being, agency and community are highly rated values, consider the practices of the green revolution to lack relevant value and they have little positive interest in the knowledge that informs them. They do not doubt that the green revolution is informed (at least in the short run) by soundly accepted theories, but they query the significance of the theories because they do not encapsulate possibilities relevant to the realization of their fundamental values. They are interested in the lost possibilities of the green revolution, and how they might be brought to realization; and they wish to nurture the social values of social and ecological stability. Research, that might serve their interests, aims to study natural objects explicitly in terms of their relations to the social and ecological orders (and not simply as objects whose possibilities are generated from the underlying order where their role in the social order is only implicit), thus poses a question like: How can we produce wheat so that all the people in a given region will gain access to a well-balanced diet in a context that enhances local agency and sustains the environment? It, in contrast to the question posed at the outset of this section: "How can we maximize wheat production under 'optimal' material conditions?", does not abstract from the conditions of daily life and experience and prevailing practical activities of a region, and it does not presume that questions of social order are subordinate to the implementation of novel controls. It does not consider the biology, ecology and sociology (or production and distribution) separately, and it locates questions about crop yields among questions of the following kinds: What are the socio-economic conditions and social effects of agricultural production? Who controls the product? What use is made of it? How is it distributed? How do the socio-economic conditions of production affect those of distribution, and *vice versa*? What are the effects on health and ecology? Thus, crop yields are investigated not only as functions of broadly materialist variables, but also of the social and other variables of which the materialist variables are themselves a function. This approach turns attention to the local and the particular: to local soil conditions, strains of seeds, ecologically sound methods, availability of "natural" pest controls, traditional practices; and to local socio-economic relations, needs, aspirations and histories. These are the kinds of matters that must be investigated if it is sought to reshape the world of daily life and experience so that control ceases to be hegemonic, but contained by the exigencies of the values highlighted by authentic development.

Some researchers connected with grassroots movements have proposed that relevant understanding can, in part, be gained from improvements (based in empirical research) of traditional practices and understanding – forms of research which follow strategies that do not abstract phenomena from the ecological and social relations that they exhibit.[11] They challenge the claim to possess a monopoly on knowledge (soundly accepted, systematic empirical knowledge of material and biological objects) often made on behalf of materialist science, by pointing to the empirically vindicated strengths of local, traditional, people's knowledge which, because of its locality, assumes numerous, diverse forms. Informed by such knowledge, traditional farming has developed, in some cases, practices that are ecologically sound (maintaining, for example, soil that has remained fertile for millennia, and pest and disease controls that function through appropriate arrangements and combinations of crops), selection processes that have generated a richly diverse gene stock, and modes of social organization in harmony with natural processes.

Aware that in dealing with these matters (in medicine as well as agriculture) one can be tempted to lapse into nostalgic romanticism or "new age" enthusiasms, I emphasize that "fitting with tradition" is a cultural, not a cognitive value. A "privileging" of knowledge gained under the materialist strategies should not be replaced with a generalized and uncritical privileging of traditional knowledge forms and claims. Regardless of strategies deployed, knowledge claims are to be assessed in virtue of how well they manifest the cognitive values, not just in virtue of their potential significance for an adopted value complex. A cultural value can motivate adopting certain strategies, but cannot ensure that the world will be amenable to grasp under these strategies. My interest is empirically grounded. First, there is the empirical record that current development practices are failing to satisfy the basic needs, to cultivate the human capacities, and to respect the human rights of vast numbers of impoverished peoples. Second, the references (Note 10) provide evidence that, *at least under some cultural conditions and in some locales*, traditional, ecologically sound and socially strengthening agricultural technologies are potentially sustainable (though clearly they cannot be developed further, and may not be able to survive, under the hegemony of policies of modernizing development). Grasping and deepening the understanding that underlies these technologies, of "agroecology," and separating out that which is empirically sound from that which is not, has become a major concern for those attempting to implement conceptions of development that are opposed to modernizing development. They push for empirical investigation of the "lost possibilities" question. They oppose both presuming that there are significant lost possibilities, and laying aside or dismissing the question because of its inconvenience in the light of those policies that foster the spread of agricultural practices that are implicated in biotechnology and expanding agricultural exports, and that further the manifestation of the modern values of control. The mode of research that they advocate does not dispense with materialist understanding, but rather proposes a dialectical interaction between traditional and materialist approaches, with the interest of materialist results subordinated to their relationship with the social values of the alternative conceptions of "development."

Shiva has referred to the relevant traditional knowledge as dealing with "preserving and building on nature's processes and nature's patterns," with "repairing nature's cycles and working in partnership with nature's processes," and with "subtle balances within the plant and invisible relationships of the plant to its environment."[12] This knowledge concerns the relationships of plants (and animals) with local physical conditions, with other biological organisms, and with people and social organization. Clearly, over the centuries, indigenous seeds (obtaining from free pollination in open fields) have been improved as a result of the selection practices of local farmers. Their knowledge can become an object of systematic investigation in which it is articulated, systematized, empirically tested and further improved. Traditional agricultural knowledge abstracts from the molecular structures and "internal metabolism" (Kenney 1986) of plants, and so cannot by itself encapsulate all the possibilities identified in research under the materialist strategies. Nevertheless, it has produced the genetic materials, which are the fundamental prerequisite of all research with hybrid and genetically engineered seeds. It is, thus, a body of knowledge with impressive empirically vindicated credentials. There is no clear reason why it should be incapable of further development.

Paradoxically, modern scientific research both presupposes the results of traditional knowledge practices, and acts to destroy them or at least to deny their products any legitimacy as knowledge (Marglin and Marglin 1990; Shiva 1988, 1997). The alternative approach, proposed by the critics under discussion, is to develop a dialectic between traditional and materialist approaches to gaining knowledge, to pursue a version of full understanding that engages not only the scientific "expert," but also the peasant practitioner.

Dialectic of traditional knowledge and research under the materialist strategies

How might a dialectical interplay of traditional knowledge (for example, a form that has a mutually reinforcing interaction with the values of enhancing local well-being, agency and community) and materialist investigation work? Or, how might investigation be conducted under the materialist strategy, but subordinated to other values? There are relatively straightforward ways: the values might prioritize research on the social possibilities and consequences of applying theories accepted under the materialist strategies, or point materialist research in a direction so as to focus upon particular phenomena deemed of special social value. Can the biological (and the material) always be demarcated from the social in this way: where social inquiry complements and partially directs the biological inquiry, but where the biological possibilities (in the agricultural case under discussion) are discerned only in relation to the fundamental underlying genetic theory? Or sometimes may there be such a profound *interaction* between "natural" and social variables that an adequate encapsulation of possibilities could not derive from research that draws upon "adding up" results gained from the standard disciplines?[13] So, minimally, adopting an alternative strategy (if the answer to the first question is "yes") may lead to a richer interdisciplinary approach, but (if it is "no") it opens up the possibility that we may need to address certain questions in ways that cut across the standard disciplinary lines.

The discussion of these issues has entered a new stage with the advent of the biotechnology revolution. Some have argued that not only do the new biotechnologies promise the development of higher-yielding varieties of crops, but also that they will not have (and may even reverse) the negative ecological effects encountered with the green revolution. And others have suggested that they can realistically serve the interests derived from values of social justice in the impoverished countries, especially those which are the source of the greater part of the genetic resources of humankind. Consider: "As a society, I think that we would like to use our enhanced capacities for manipulating the genetic code to develop and deploy new plant varieties in ways that are economically productive, socially equitable, and ecologically benign. Will we be able to do so?" (Kloppenburg 1988: xiv). Putting the question this way accepts that the science of genetic engineering will develop of its own internal dynamic, so that we are confronted with the need to find ways to use it to serve popular interests rather than those of capital. This science is just a reality to be coped with, hopefully constructively; it frames questions about agricultural policy today. It will have some role; the question is: what role? Following this logic, the poor countries would need scientific institutions that engage in specifically targeted applied research in biotechnology, or perhaps also in focused "fundamental" research linked to those applied areas. Kloppenburg, however, is sensitive to the "lost possibilities" consideration, and to the potential for improvements in traditional technologies; and so he urges research in this area too. His solution seems to be a "both ... and ... " rather than a dialectical interaction. Basic biotechnological research is presumed to be significant, with the specific focus of its applications to be judged (and compared with traditional methods) in the light of economic, sociological and ecological considerations. But he does not evaluate the significance of the research program of biotechnology in the light of the interests of authentic development (as one seeking a form of full understanding would).

There is something appealing about the "both ... and ... " position, as it attempts to be responsive both to the realities of power and to the rights of the poor, and it needs fuller consideration. Nevertheless, unless the institutions of biotechnological research were to be radically changed and the research relocated within a program of full understanding related with the values of social justice, biotechnology will continue for the most part to serve the interests of expanding capital-intensive agriculture. Genetic engineering furthers the process of turning the seed into a commodity, since genetically engineered seeds may be patented, and the dominance by agribusiness of the market can create pressures for farmers to use the new seeds. According to international trade agreements, the use of such seeds is restricted in accordance with the laws of intellectual property rights. Given the cost of biotechnology research, it is difficult for communities and institutions in poor countries to claim such rights effectively for themselves. This is clearly disadvantageous for them. It not only entrenches dependency and inequity in trade relations, but also weakens the ability of the poor nations to utilize for their own ends one of their most valuable natural resources, the richness and variety of their genetic resources.[14]

It remains for those with the appropriate technical competence to explore the promise of any efforts to develop a dialectic between traditional knowledge and materialist science. In doing so, it should be kept in mind, paralleling my discussion of the materialist strategies (Chapter 5), that alternative strategies being adopted in view of their mutually reinforcing interactions with certain social values is neither sufficient for accepting "theories" developed under them, nor *per se* an objection to engaging in research under them – as long as the roles of the cognitive and the social values are kept separate. The role of the cognitive values is essential. Perhaps the world will not be amenable to grasp under the alternative strategies! But it is always relevant to ask, when research under certain strategies is facing difficulties, if the difficulties derive from the way the natural world is or from opposing social forces.

Working out alternative forms of understanding needs to be done subtly and realistically, with full awareness that the dominant structures of power are cast against the alternative conceptions of "development" with which they have mutually reinforcing relations, and with full awareness of the social relations of scientific research and development. In particular, it requires awareness that modern (largely materialist) science, like capital, is not under the control of agencies in the impoverished countries, and that it is dialectically linked both with the modern values of control and with capital. While some of the possibilities, uncovered by materialist science, may be pertinent to the interests of authentic development, its general impact can be expected to be linked with the programs of modernizing development. In the light of this, it will be difficult to obtain the institutional conditions in which to investigate sharply whether difficulties facing alternative projects stem from the way the natural world is or from opposing social forces. It is frequently said today that entertaining possibilities that cannot be realized within the structures of modernizing development is "unrealistic" (Lacey 1997c). If so, is the source of the lack of "realism" from the natural world, or from social forces (including the use of power)?

A "grassroots empowerment" approach

Drawing upon the discussions of the seed and of authentic development, we may entertain as the objective of an approach to science (O_2) one that adopts strategies – involving a dialectical interplay between the methods of gaining traditional knowledge and research conducted under the materialist strategies – that enable us to identify those possibilities whose realization would further the interests of authentic development and the fuller manifestation of its prioritized value, enhancing local well-being, agency and community. The strategies to be followed (and their variation with locale) remain to be specified in detail, but the approaches in which they are adopted are effectively equivalent to a "grassroots empowerment" approach:

O_2' The objective of a grassroots empowerment approach to science is to encapsulate (reliably) the possibilities of domains of objects (for example, objects relevant to agricultural practices) that are of potential value for projects responsive to the value of enhancing local well-being, agency and community, to discover means to realize some of the hitherto unrealized ones, and to preserve those already realized.

Clearly if one's value complex includes the modern values of control, then one will adopt O_1. The point of O_2 will be apparent to those whose social values make them critics of the modern values of control, who maintain that the embodiments of these values require practices and institutions that can only be maintained in an economic order which inherently has undesirable consequences, such as social and ecological devastation, unacceptable inequalities, patriarchal relations, alienated labor, or class-based relations of domination. Thus, the critics will ask: How shall we interact with nature so as to serve the coming to be of an alternative social order (authentic development), which (for example) enables grassroots empowerment to be enhanced (or which embodies some different view of social justice)? For this end, what will be the characteristic way (ways) of interacting with nature? What kind of strategies (alternative to the materialist strategies, or to which the materialist strategies are subordinate) should be brought to bear in order to gain empirically grounded knowledge that would serve that end? As we have seen, there may be interesting mixtures of pre-modern and (post)modern answers to these questions.

9   A feminist approach

There is no objection (in principle) to identifying and engaging in approaches to systematic empirical inquiry that deploy strategies distinct from the materialist strategies (Chapter 5); and I have explored the motivation, prospects and implications of one alternative, the "grassroots empowerment" approach (Chapter 8). In this chapter, I explore another alternative approach, one whose strategies interact in mutually reinforcing ways with "feminist" values. Again, provided that the distinction of the roles of the values and the cognitive values is respected, this alternative need not involve bringing values to bear on scientific claims in any way that is logically different from that involved when the materialist strategies are adopted.

THE VERY IDEA OF A FEMINIST APPROACH

Feminism is a political movement with implications for restructuring the whole gamut of social institutions. Few seriously query the propriety of feminism to have an interest in adopting a critical stance aiming to uncover biases that may be at play in current scientific practices, in advocating that more research be conducted on specifically "women's problems," for example, in the health sciences or in investigating possible barriers to the admission and advancement of women in scientific institutions and proposing ways to eliminate them. There is strong resistance, however, to the idea that there may be a feminist approach to science which draws positive direction from feminist values. Geertz expresses the resistance clearly:

> The worry is ... that the autonomy of science, its freedom, vigor, authority, and effectiveness will be undermined by the subjection of it to a moral and political program – the social empowerment of women

– external to its purposes … [namely] the knowledge-seeking ones of
science, the no-less-impassioned effort to understand the world as it,
free of wishing, "really is."

(Geertz: quoted in Lloyd 1996)

The worry is that any scientific practice that bears the label "feminist" will
not be value-free in ways widely thought to be essential, at least as
aspirations, to scientific practices. It concerns the very idea of a feminist
approach to science, not the results of its actual practice and not objections
to feminist values *per se*.

This worry is misguided. The *autonomy* to which Geertz appeals does not
hold even of research conducted under the materialist strategies (Chapter
10). A feminist approach to science may extend "knowledge-seeking"
practices into hitherto neglected domains, though how fully it can be
implemented remains a matter for both further empirical investigation and
political activity, and the significance of such extensions can be contested.
While a strategy may be adopted because it interacts in mutually
reinforcing ways with particular values (whether the modern values of
control or, for example, feminist ones), adopting the strategy does (and can)
not commit one to accept any concrete theory; and, in the long run, a
strategy ceases to be (cognitively) worthy of adoption if it fails to generate
theories that become accepted (Chapter 10). I belabor this point because
barriers to its recognition are strong, for in the mainstream scientific
community (and much of the philosophical) it is rarely seriously queried
that *autonomy* is in fact realized to a tolerable approximation. This view, in
turn, both gains support from, and reinforces, the further view that theories
consolidated under the materialist strategy broadly characterize the world
as it is rather than that primarily they characterize objects grasped from the
stance of control. It is the grip of materialism that puts all incipient
alternatives at an additional disadvantage. They, but not research under the
materialist strategies, have to defend themselves in face of the charge of
the "intrusion of politics." In view of the way in which the modern values
of control are among the values highly expressed in the dominant
contemporary economic, social and political institutions, so that their claim
appears to be universal, this may be inevitable. Nevertheless, there is no
sound argument for singling out the alternatives in this way. That they are
so singled out, antecedent to the outcomes of research under the
alternative strategies, I guess we can say, is "ideological."

LONGINO'S APPROACH

The idea of a feminist approach to science, which I discuss in this section, derives from reflecting on recent writings of Helen E. Longino (especially Longino 1990).[1] Longino states her version of "feminism" as follows:

> Feminism ... is at its core in part about the expression of human potentiality. When feminists talk of breaking out and do break out of socially prescribed sex roles, when feminists criticize the institutions of domination, we are thereby insisting on the capacity of humans – male and female – to act on perceptions of self and society and to act to bring about changes in self and society on the basis of these perceptions.
>
> (Longino 1990: 190)

Lying behind feminism, so understood, is a conception of human nature: human beings have the capacity to act informed by their own values in the light of their assessments of current realities, and to act efficaciously to "bring about changes in self and society on the basis of those perceptions." Human beings are agents, with "capacities for self-consciousness, self-reflection, and self-determination," and whose intentional states are efficacious. This conception of human nature serves to ground such prioritized values as "liberty, autonomy and responsibility." It also is seen as "partly a validation of our ... subjective experience of thought, deliberation, and choice."[2]

Agency is a human capacity. Its exercise can gain positive conditions from one's relations with others and one's places in social institutions, or be diminished by them. Historically, diminished agency has been common, but affecting various groups of people differently. In addition to the conception of human nature, Longino's feminism endorses the possibility of enhancing agency where it has been diminished, a possibility whose realization itself depends upon concerted action which will depend upon identifying and eliminating the causes of diminished agency.[3] Expanding the exercise of agency becomes the central objective of a political movement. What kind of approach to science might serve this objective? Or, as Anderson puts it, what are the "scientific practices, which incorporate a commitment to the liberation of women and the social and political equality of all persons" (Anderson 1995a: 51)? Or, noting that factors pertaining to gender are among the causes of fundamental inequalities, what kind of approach to science might serve the narrower objective that sometimes Longino focuses upon: "to reveal gender," that is, to reveal the action of "an asymmetric power relation that both conceals and suppresses the independent activity of those gendered female ... , [a] relation ... sustained by social institutions and symbolic practices ... itself made invisible as a relation of power by ... naturalizing models in the life and behavioral sciences of sex and gender differences" (Longino 1996: 50).

Feminism emphasizes enhancing human agency, and so it subordinates the modern values of control. In principle, feminist values can be linked dialectically with alternative strategies for pursuing the objective of encapsulating the possibilities of domains of phenomena. We can see this best by developing further a contrast with the materialist strategies. When I considered the reasons for adopting the Galilean/Baconian approach (O_1/O_1'), the focus of attention was human interactions with material objects. There (Chapter 6), in effect, a question like: "How must material objects be thought of, if control is to become the characteristic human stance towards them?" was posed. The answer – summarily: as grasped under the materialist strategies – frames an approach which produces theories that have come to inform practices which bring about a fuller manifestation of control as a social value within the predominant social institutions. More fully manifested control is evident in the creation and multiplication of technological objects, which become the center and the cutting edge of the productive process; and so it has implications concerning the shape, possibilities and structuring of human lives. Understanding of the material workings of technological objects is gained by adopting O_1. This form of understanding treats these objects in abstraction from their social (and ecological) contexts. It attends to their material products and to the processes underlying their production. But it lacks the resources to offer understanding of the social institutions of production, distribution and use, the values and interests they represent, and disparities they may introduce and require concerning enhanced or diminished human agency.

Materialist understanding does not answer: "What do people become like (and with what kinds of variation) when understanding, gained under the materialist strategies, is applied within modern institutions?" The natural science is dissociated from the sociology. This seems to be a corollary of focusing almost exclusive attention on the question posed in the previous paragraph – where we consider the modern values of control in abstraction from the conditions required for their becoming highly manifested in social institutions, as if their universality or desirability could be taken for granted. Control can indeed be distinguished from such other values as meeting people's material needs, profit, efficiency, serving the free enterprise system, or developing socialist or feminist consciousness. But within actual institutions, it will be manifested along with the high manifestation of some and the low manifestation of other values. How one ranks control as a social value, thus, can be a function of rankings of the other social values from which (contextually, historically) it cannot be separated.

Now, consider these questions:

1   How must human beings be thought of if feminist objectives are to be furthered?
2   What kinds of knowledge could inform the furtherance of feminist objectives?

In posing questions like these first, the feminist approach under discussion is the opposite of the one I attributed to materialism. It *begins* with what human beings are, not with our relations with material objects and not with what material objects are. Suppose that those stressing the modern values of control also asked: How must we conceive of human beings if they are successfully to adopt control as the characteristic stance towards material objects; and if they are to apply scientific knowledge to further such control?[4] It seems that two answers would have to be given. First, human beings are agents, capable of acting in the light of their knowledge for their own ends. Second, human beings are adaptable to the roles (or absence of roles) required of them in relation to technological objects, especially to the means of production. As general answers they are incompatible. Yet both are needed. Some human beings will have to be one way, some the other; and at least the second involves diminished agency.[5] Feminism challenges this bifurcation (in part, because it is often drawn along gender lines), emphasizing the universality of agency, and questioning those approaches to science – in psychology and sociology – that attempt to ground the bifurcation in inherited natural differences.

Answers to the two questions just raised highlight agency as the distinctive human capacity shared by all human beings, but diminished in its exercise under some social conditions. Then, human beings as agents – and the conditions under which agency is enhanced or diminished – become the object of inquiry; and attempts will be made to identify strategies for an approach (O_3) that is effectively equivalent to:

O_3' The objective of a feminist approach to science is to identify (reliably) the possibilities open to human agency, the conditions for its enhancement and diminishment, and to discover means for bringing about more of the enhancing conditions.

I will make no effort to specify all the implications of following O_3,[6] but only those fairly directly connected with human agency and especially with cognitive abilities and activities, areas of investigation proper to the biological, psychological and social sciences. I leave it an open question whether a version of O_3 can be followed in the physical sciences. It has become commonplace to affirm that feminist insights (both critical and positive) will be confined to the "soft" sciences. If so, so be it. But I am not so sure (cf. Tiles 1987), for one may ask: How must we think of material things, and what must be our characteristic stance towards them, if we wish to enhance the agency of everyone maximally? Different cultures and locales may produce richly different answers to this question.

Constraints on theories

Following O_3 involves adopting strategies that put constraints on admissible forms of understanding of human phenomena.[7] The first constraint is that intentional explanations have primacy in the domain of human behavior. Agency is unintelligible apart from the causal efficacy of an agent's beliefs and desires. So, the primary categories of behavioral explanation will be belief, values, desire, intention and the like, rather than posits of underlying law, process and structure. Seldom will such explanations be articulated with the kind of formal (deductive, mathematically articulated) organization present in materialist theories. Usually they will be articulated in the form of narratives enriched with situation-bounded regularities (Lacey and Schwartz 1986, 1987). The second constraint is that diminished agency will typically be interpreted as socially produced – though, in special cases, it will be considered as physiologically or psychologically produced – and sometimes social accounts are supplemented by psychoanalytic ones (Keller 1982); and (it seems to me) it will tend to be reducible to behavior that more completely fits law-like regularities. Identifying the social/historical bounds of such regularities will constitute an important research item. These constraints, as well as the dialectically related feminist values, draw upon the view that human action, since it is intentional, does not reduce to behavior, as understood in "scientific" psychology (psychology conducted under materialist strategies: Schwartz and Lacey 1982).

Strategies in which the hypotheses that can be entertained are bound to consistency with the intentionality of human agency are incompatible with materialist strategies, for action cannot be represented simultaneously as intentional and lawful (Donagan 1987; Lacey 1996).[8] Thus a presupposition of the feminist value complex is inconsistent with numerous hypotheses about human behavior entertained under materialist strategies, for example, those versions of materialist strategies that restrict the kinds of variables that are admissible in laws (and thus in underlying structures and processes), and so aim to explain behavior in terms of its lawful relations with genes, fetal hormonal phenomena, settled brain states or environmental contingencies. Illustrating this Longino discusses a strategy in behavioral endocrinology that uses "the linear-hormonal model" (LHM: hormone – brain-organization – behavior),[9] in which sex differences in a variety of behaviors (including performance in some mathematical tests) are lawfully attributed to differences in brain organization, which are themselves lawfully attributed to differential roles of gonadal hormones in fetal development.

Concerning research under LHM, Longino says: "Our political commitments ... presuppose a certain understanding of human action, so that when faced with a conflict between these commitments and a particular model of brain–behavior relationships we allow the political commitments to guide the choice" (Longino 1990: 191). "Guide the choice" to what? To rejecting out of hand any theory developed with LHM? But that would be an instance of fact being inferred from value, and of "wishing" (Geertz) that the world be consistent with the presuppositions of her feminist values.[10] On closer inspection, however, it is clear that the choice Longino is referring to is to engage in research under the feminist strategies rather than to deploy LHM. This choice cannot properly be dismissed simply as a matter of "wishing" unless theories developed with LHM in fact manifest the cognitive values to a high degree of some relevant domains of phenomena. Longino argues that they do not, and (in effect) that they only appear to do so in the light of assuming that there are essential brain differences between males and females that account for a great variety of behavioral differences. This assumption is not contained in a theory soundly accepted in accordance with *impartiality*, and holding it is explicable only in terms of its being a presupposition of value complexes which legitimate sex differences (and male superiority) in a wide range of social roles. It follows that – as the state of evidence now stands – accepting theories developed with LHM involves the play of a value (a "sexist" value: preserve sex differences in social roles) *alongside* the cognitive values rather than *prior to* their play (in the course of adopting a strategy).

The different roles of the social and cognitive values are crucial. There are no *logical* and *methodological* (as distinct from moral) objections to adopting the strategies of LHM on the ground of their mutually reinforcing interactions with value complexes that legitimate widespread gender differences in social roles. Usually, however, those who adopt such value complexes maintain that they do so in large part because "natural" gender differences have been established rather than that they are attempting empirically to discover "natural" gender differences for the sake of confirming the presuppositions of their value complexes. No one, I suspect, adopts LHM explicitly because it can be expected to provide support for "sexist" values – hence my conclusion that, where theories developed with the model are accepted, "sexist" values are in play alongside the cognitive values. Even so, it is not "bad" science *per se* to engage in research that deploys LHM (Nelson 1995, 1996), especially if it is the only or the principal available option in an area of investigation, as long as one only accepts its theoretical products if they manifest the cognitive values highly according to the highest available standards. Furthermore, the unfolding of research using the model may increase the severity of testing theories developed under any rival strategies, and this fact may provide sufficient reason to support such research as one among several approaches. While it is proper to adopt strategies whose theoretical products can be expected to be significant for one's moral projects, it is also important for the scientific community to ensure the developed theories are subject to testing with sufficient severity.

The criticism just made of accepting theories developed with LHM is *logically* (but not *causally*) independent of the "political commitment" to which Longino refers. Although it originated in the course of inquiry under O_3, its logical and methodological force does not depend on sharing the presuppositions and values lying behind O_3. Having made the criticism, then (logically) the choice is made to follow O_3, to follow a research approach whose strategies are dialectically linked with feminist values, *and* which runs counter to an approach in which at best theories with low manifestations of the cognitive values have been produced. Making that choice, of course, does not provide a ground for the judgment that research deploying LHM will not be empirically fruitful, though it does for the judgment that the products of that research will probably not be very significant for feminist value complexes (Anderson 1995b). In choosing to follow O_3, one identifies human agency as the primary object of inquiry, and one is primed to query alleged limits to its possible expansion based on gender, race and other such differences. As the criticism of LHM makes clear, following O_3 also contributes to "making gender visible," by bringing to attention the play of "sexist" assumptions alongside the cognitive values in some scientific practices.

Positively, for Longino, the feminist strategy can encourage research on models of the brain that conflict with LHM (such as Edelman's "selectionist model"). A model must be sufficiently complex to be consistent with the intentionality of action – "a model [that] allows not only for the interaction of physiological and environmental factors, but also for the interaction of those with a continuously self-modifying, self-representational (and self-organizing) central processing system." More generally, the assumptions tend to constrain theories in the direction of "complexity, ontological heterogeneity, interaction and holism (non-reductionism)." For Longino feminist strategies involve these constraints, not because there is an alleged special "female sensibility or cognitive temperament" that might value highly such characteristics or claim special insight that the world is this way, but because human agency displays these characteristics, and cannot be readily recognized and may easily be diminished when these types of characteristics are not highlighted.[11]

Selection of data

Recall that a strategy both constrains the class of potentially acceptable theories, and selects the kinds of empirical data deemed relevant for the appraisal of theories that meet the constraints. Recall, too (Chapter 7), that the act of observation does not uniquely determine the empirical data, and that there is no set of lexical categories especially apt for reporting them. Furthermore, objects of experience are not limited to phenomenal or subjective states; typically they include intersubjectively observable states of affairs that involve material objects and (often) the actions and interactions of human beings. The relevant lexicon for reporting empirical data depends on the objects of experience that are of interest for the investigation at hand. So adopting the strategies that define O_3 involves selecting certain kinds of data as especially pertinent for the investigation of agency, so that they can be brought to bear on theories that meet the constraints. Agents act to realize their desires (shaped by their values) in the light of their beliefs. Reflecting this, the data include reports of action in intentional idiom. Actions are observable (Donagan 1987), and the language of empirical reports of actions need not abstract from the intentionality involved. One observes (for example) that the experimenter switched on the machine and read the meter, both actions: neither bodily movements nor the environmental effects of bodily movements (though generally there is no action without bodily movement and its environmental effects), neither reflexes nor operants. Action is not behavior, as pointed out on p. 205, an object (characterizable completely in non-intentional idiom) which follows lawfully from the action of, for example, genetic, environmental and neuro-physiological variables. Actions will be observed along with their verbal accompaniments (themselves actions), as well as numerous contextual factors including the social interactions of which they are part, so that where actions do not follow from a person's authentic values and beliefs (when agency is diminished), it becomes possible to explore the social factors that may be among the social causes of that diminishment.

LHM cannot account for action; lacking the necessary lexical resources, it cannot fit the kinds of data just described. It can only account for behavior, its variations and its distribution across groups. Not being able to account for action is the ground for a further criticism of LHM, different from that made earlier and one that cuts deeper. The earlier criticism simply cleared obstacles to adopting the feminist strategies. It maintained that theories developed in accord with LHM do not manifest the cognitive values highly, so that the presuppositions of feminist value complexes are not inconsistent with theories accepted in accordance with *impartiality*, and that the belief that any of these theories are soundly acceptable comes from (implicit) links with "sexist" values. As I pointed out, this criticism poses no logical and methodological obstacles to continuing to engage in research under LHM (but not exclusively), and the results of such research can be evaluated in the light of the data and the cognitive values. Despite the alleged link with "sexist" values one might support engaging in research under LHM on the grounds that its strategies instantiate the materialist strategies and that it is a fruitful research program, the kind of research that has proved itself in many fields to be very productive. Especially where the conditions for conducting it are in place, we would expect this line of reasoning to be quite attractive. But it does not undermine the reasons proposed for choosing to adopt the feminist strategies for investigating human cognitive powers and their exercise. Each line of reasoning has its appeal; but neither, in actual fact, is generally compelling, leaving, it appears, at the present time – so far as logical and methodological considerations are concerned – for individual investigators an open choice regarding which strategies to adopt, and for the scientific community an interest in all open options actively being adopted by some of its members.

"Open choice of strategy" expresses, however, not the last word, but rather an interim tactic to be deployed pending further developments of investigations under the various strategies. For dealing with human cognitive abilities and their exercise, feminist strategies and the strategies of LHM (or any variant of the materialist strategies, at least as currently developed) are incompatible and produce incommensurable theories. The latter strategies do not possess the conceptual resources to investigate empirically the presupposition of feminist value complexes about the centrality of human agency. Does following O_3 rather than O_1 (one of its variants, for example, LHM) express just an interest in different possibilities, perhaps in those of action rather than of behavior – an interest in different 'worlds' (Chapter 7)? If so, each could proceed unimpeded by the other (except insofar as they might compete for resources) and, if successful, encapsulate progressively more of the possibilities respectively of interest. Up to a point this is a useful way to look at the relationship between the two approaches, to consider them as dealing with different 'worlds.'

On the other hand, they compete to produce theories that apply in the social "world" of daily life and experience, both to explain phenomena encountered and characterized in the course of common social practices, and to come to inform action in this realm. Given the ubiquity of intentional idiom in the characterization of most common social practices, *prima facie* any approach (for example, feminist) that admits a role for intentional categories has greater explanatory power concerning human phenomena of salience in them (for example, the phenomenon of engaging in systematic empirical inquiry itself: Lacey and Schwartz 1987) than an approach defined by materialist strategies. For present purposes I will not develop this point. Rather I emphasize two others: (1) that proper testing of some theories (of domains that include characteristic human phenomena of daily life and experience) developed following O_1 requires reference to, and comparison with, results obtained using different strategies; and (2) that in order to legitimate practical applications of the products of following O_1, one needs to show that there are no undesirable consequences, for example, impairments of human agency (Lacey 1979) or lost positive possibilities (Chapter 8), but the research to show such things cannot be conducted under strategies of O_1 alone.[12]

Concerning (1) consider, for example, the hypothesis (h) that mathematical abilities in boys are greater than those in girls. This might perhaps be thought to be established, as part of a theory utilizing LHM that relates mathematical ability (and its manifestations in behavior) linearly to brain organization, on the evidence that boys perform on average significantly higher than girls in an array of standardized mathematics tests. Longino points out that for the cited evidence to count in favor of this hypothesis, rather than (h') that boys and girls are exposed differentially to a variety of social factors, one would also need evidence supporting such assumptions as: the data have been analyzed so that all relevant social factors have been controlled for; test performance is reliably an indicator of inherent ability rather than of acquired knowledge or learned abilities; there is one form in which mathematical ability is expressed and that form is expressed in performance in tests of the kind given in the studies; and that the content of a problem has no bearing on its formal properties or on an individual's grasp of those properties (Longino 1990: 126–7). A theory containing h manifests the cognitive values highly, according to available standards (Chapter 3), only if it manifests them more highly than competitors containing h', but those competitors are not developed under LHM (and closely related approaches). Where proper attention is not paid to this standard, it becomes difficult to identify undiscerned or unacknowledged roles that values may be playing alongside the cognitive values. Thus the proper testing of h can be carried out only in a context where approaches other than O_1, which permit empirical investigation of h' (for example, O_3), are also followed.

Concerning (2) suppose it were suggested that a sensible practical application of h would be, for example, to put more resources into mathematics education for boys than for girls so as not to waste resources teaching mathematics to those with lesser abilities. Now, apart from the questions of the adequacy of the testing of h, of whether the intended goal of the application is desirable and of whether there might not be better applications, the moral legitimacy of the proposal depends on there being no undesirable side-effects of implementing the application. The very application, for example, might reinforce other social factors that "really" account for the differences of performance, and itself become one of them (Schwartz 1997) by for instance not providing the conditions that enable the ability to develop and to be expressed (Lacey 1979). This is not an idle possibility when one considers a child's performance on a text to consist of a series of actions, each responsive to a variety of the child's desires and beliefs. Again, following only approach O_1 (any version) is inadequate to the task of investigating such conjectured potential side-effects. When the materialist strategies are followed exclusively important "lost possibilities" questions always remain outside of empirical purview. They limit what can come into view, and in a way that is likely to be detrimental to the furtherance of feminist values.

Thus, the "open choice of strategy" argument breaks down when questions of rigorous testing and of application are addressed in addition to those of the classes of possibilities of interest. Research on human cognitive abilities and their exercise, conducted solely under the materialist strategies, cannot adequately address these further questions. It needs to be complemented minimally by research conducted under strategies that admit roles for intentionality and for social causation (for example, those of O_3). A similar conclusion (that it needs to be complemented by research under conflicting strategies) also holds for O_3, but unlike the one for O_1, which is often resisted, it has little practical punch. Given the current modes of functioning of scientific institutions, those who adopt feminist strategies cannot avoid having to "legitimate" their approach in the face of those who adopt materialist strategies, and thus critically testing their own products against those developed under the materialist strategies (for example, LHM). Testing across strategies, analyzing competitors in the context of a form of full understanding (because it wants to "uncover gender"), and critical self scrutiny are all integral parts of a feminist approach. Nevertheless, it is important to emphasize them given the objective of gaining theories (posits) that manifest the cognitive values to a high degree as assessed by the most rigorous available standards, for strong presuppositions about agency lie behind feminist value complexes. *Prima facie*, approaches such as those of radical behaviorism, LHM and sociobiology have many obstacles to overcome before their interpretations (reductions or replacements) of agency and thus common social phenomena become sufficiently detailed to compete with intentional accounts.[13] On the other hand, detailed empirical charting of contemporary social phenomena will certainly display (at least in some groups, contexts and institutions) a wide variety of law-like (or, if not strictly law-like, statistically stable) regularities. Some of these uncontroversially reflect deliberately adopted social conventions, but with others it cannot be settled a priori whether (and in what cases) they are genuinely lawful (derived from underlying law, structure and process) or historically variable regularities, explicable in terms of the boundary conditions and practices of historically specific institutions that bring about diminished agency (Lacey and Schwartz 1986, 1987).

Hence I emphasize (though, as mentioned, it is routinely recognized by its followers) that the following of O_3 needs to be complemented with research under the materialist strategies. Thus, following O_3 could (in principle) have the consequence that theories – that may include h – developed under a version of O_1 come to be accepted of human cognitive abilities and their exercise.[14] But following O_1 in the customary way does not even lead to entertaining theories which deploy intentional categories. I said that the criticism of LHM, implied in pointing out that it cannot deal with action, cuts deeper than the earlier ones designed to carve out some space for research under feminist strategies to proceed. It cuts deeper by moving beyond supporting that O_3 is just one legitimate option – in mutually supporting interaction with feminist values – among potentially many, to making clear that O_3 (or close variant) is essential – *for cognitive reasons* – for appraising theories of human cognitive abilities and their exercise and legitimating their practical application.

Comparing empirical adequacy across strategies

The preceding argument follows from the claim that LHM and, more generally, any theories (T_M) generated under the materialist strategies cannot fit certain empirical data: reports of observed actions. Does anything follow about the limitations of the theories (T_F) developed under O_3 from the apparently symmetrical claim that the theories it produces cannot fit another kind of empirical data: reports of behavior and its relations with, for example, environmental factors, most importantly experimental data involving these phenomena?

This raises an important question about what constitutes the empirical adequacy of a theory. What data should a theory be expected to fit (Chapter 10)? A theory is developed under a particular strategy. In general it cannot be expected to fit (to put unificatory or predictive order into, as distinct from merely to be consistent with) data generated from an incompatible strategy. An *acceptable* theory (of a specified domain) should fit the data selected by the strategy under which it was developed; a *significant* theory should also apply to salient phenomena located in the "world" of daily life and experience, so that the data it fits include descriptions (stated in its strategy's lexicon) of phenomena in this "world." These descriptions are not necessarily or generally those commonly deployed within the practices in the "world" that encompass these phenomena (cf. the discussion of the child's swing in Chapter 7). From different strategies we may get different and incompatible descriptions (cf. impeded free fall, damped pendulum) of the same phenomenon (the swing), all produced for the sake of explaining it and identifying its possibilities. The most adequate of these descriptions will be the one fitted by a theory that manifests the cognitive values highly; then we have grounds to affirm that the phenomenon as encountered in the "world" of daily life and experience is identical to that characterized using the lexicon of a particular strategy.

Numerous human phenomena in the "world" are described with the categories of action and intentionality, and these same categories are central under feminist strategies. When I say that theories, T_M, cannot account for action, my objection is not that they cannot fit data generated in the course of following O_3, but that the descriptions that *they* provide of characteristic phenomena of the "world," that are routinely described in intentional idiom, lack the fine-grained detail of their intentional counterparts and that thus the theories they fit are weak (comparatively) in explaining the details of the phenomena, in encompassing their possibilities, and in anticipating novel possibilities. T_M perhaps account well enough for those actions that fall under law-like (or statistically stable) regularities; where the phenomena they apply to also include experimental phenomena,[15] a case can be made that the theories are both acceptable and significant. But the strategies under which they have been developed lack the categories to define the bounds of their application. Then, if the regular phenomena are considered also characteristic phenomena, it will be held that the theory applies in a wide-ranging way across human phenomena with only the details to be worked out for the more "complex" behaviors (such as those involved in the conduct of scientific research, or a conversation entered into before making a decision). Then, action would be understood by reference to exemplary experimental (or mathematically modeled) phenomena (just as the motions of the swing became understood with reference to the pendulum; at one and the same time an experimental artifact and a mathematical idealization).

Are not T_F seriously empirically inadequate since they do not fit the data gathered in these experimental investigations? Well, in line with the discussion about which data T_M should fit, they are not, since the strategies under which they were developed provide no (central) role for experimental data of behavioral phenomena. T_F would be seriously challenged only if theories (instances of T_M) that fitted both the experimental and characteristic human phenomena manifested the cognitive values highly. They will not be challenged if evidence can be offered that the limits of application of T_M are bounded and do not include characteristic phenomena of agency. This can best be done by proposing an explicit account of the alleged bounds in some T_F and offering socio-historical evidence that they are not gone beyond.[16] Of course, to the extent that such evidence has not been obtained, the more the partial grounding of T_M in experiment will make them compelling, especially if the reach of application of T_M progressively expands as they develop. T_F does not have to fit the experimental data that support T_M to be empirically adequate, but it does have to fit empirical data (reported with the use of intentional and social/historical categories) about T_M and the reach of their applications (see Chapter 10 for a more general analysis).

Feminist approaches and impartiality

A feminist approach to science aims to produce acceptable theories (posits), grounded in the play of the cognitive values held to high standards. It is not simply a critical perspective aiming to uncover bias in mainstream science, and not simply a world-view or moral vision. It does not want to replace rationally acceptable theories with "wishings" about the way the world is. It does criticize certain scientific claims for reflecting bias, but the more fundamental criticism is that they are false, weakly manifest the cognitive values of relevant domains of phenomena, and ignore (or not permit the development of) more promising approaches. Considerations based on the cognitive values are the most fundamental ones in these critiques. Pointing to the role of bias in the support of a theory indicates that the theory is not acceptable in accord with *impartiality*, and it indicates avenues to follow for its more stringent testing – test it against competitors that are free from *that* bias. All of this makes sense because we expect our research projects to leave a residue in the stock of knowledge. If we did not expect this, we would just have the back and forth play of biases, with only power to settle the matter.

Posits soundly accepted in accordance with *impartiality* are important for feminist science also for the sake of informing feminist projects. Action is more likely to be effective if based on such posits, than if based on weakly confirmed conjectures, and much more so than if based on mere "wishing" (Geertz).[17] Moreover, the rational claims of such posits do not rest upon accepting the feminist value commitments. So feminist science aims to find out what, "free of wishing," *is* the case about human agency, its material necessary conditions, the causes of its diminution, the possibilities and means for expanding its exercise, and the barriers to it expanding. Once again, to count as soundly accepted, posits on these matters must display high manifestations of all the cognitive values measured against the stated standards.

It does not require commitment to feminist values to recognize that certain posits have (or have not) been accepted in accordance with *impartiality*, just as one does not have to be committed to the modern values of control to recognize that numerous judgments made in materialist science accord with *impartiality*. Anyone, in principle, regardless of his or her value commitments, can recognize that a theory (developed under a particular strategy) manifests the cognitive values highly. That, I suggest, is a condition for the very possibility of critique of an approach to research In practice, however, since such recognition involves certain capabilities and skills, and understanding of the meanings of terms involved, and these are often acquired only from participating in a certain kind of life or even moral project, many outside of the practice (be it feminist or materialist) may not always be able to recognize which judgments accord with *impartiality*. I take the extent of this kind of *de facto* incommensurability to be a matter for empirical investigation, not for a priori resolution.[18]

Impartiality remains a viable and obligatory ideal of scientific practice, regardless of the approach taken. The very idea of a feminist approach to science does not challenge this; indeed it should insist on it. Geertz's worry is misplaced, as long as we separate clearly the roles of social and cognitive values. The former influence the *strategies*; the latter play their role in the assessment of concrete posits. And so the role of the social value "expansion of human agency," which links with the strategy to constrain theories in "accord with complexity, ontological heterogeneity, interaction and holism," is separate from that of cognitive values; just as the role of the modern values of control, which have dialectical links with the materialist strategies, is separate from the role of cognitive values. A feminist approach to science, of course, may be criticized: one may question the concrete outcomes of the research, including how well they reflect *impartiality*, or one may challenge feminist values themselves and thus the significance of any of the concrete outcomes. Such criticisms leave the idea of a feminist approach to science sound and intact.

QUESTIONING THE DISTINCTION BETWEEN COGNITIVE AND SOCIAL VALUES

For my defense of the legitimacy and even indispensability of a feminist approach to science (O_3) the distinction between and separation of roles of cognitive and social values is crucial. While my account of O_3 draws heavily upon the writings of Longino, it cannot be considered unambiguously an interpretation of her approach. In recent articles she questions the crucial distinction between cognitive values and (other) values. She questions whether "paradigmatic constitutive values [of scientific practices] have a solely epistemic or cognitive basis" (Longino 1995: 384), and later she entertains the more radical idea, "to cast doubt on the very idea of a cognitive value" (Longino 1996: 42). If her questioning can be upheld, the defense I have mounted collapses, not only of the place of O_3, but also of *impartiality* in general and, with the latter, all prospects of gaining significant *knowledge*.

Longino entertains a possible distinction between "constitutive" and "contextual" values of science.[19] Constitutive values are criteria for what "counts as good or acceptable scientific judgments" (Longino 1995: 353), criteria "involved in the assessment of theories, models and hypotheses, guiding their formulation, acceptance and praise, disparagement and rejection, and pursuit or abandonment" (Longino 1996: 49). Contextual values are those that influence the process of science in any other way. She does not distinguish (as I have done: Chapter 1) accepting from other positive stances one may adopt towards a theory, for example: provisional entertainment, commitment to explore, application in practice; she does not distinguish judgments of acceptance and significance; and she does not make explicit that theories are accepted of particular domains. Then, constitutive values include all those that serve as criteria of scientific judgment when adopting any one of these stances, thus including judgments of both acceptance and significance. Constitutive values, therefore, include both cognitive values and at least those social values that interact with the adoption of strategies in mutually reinforcing ways. For her, cognitive values are "characteristics enhancing the likelihood of the truth of a theory or hypothesis" (Longino 1995: 583), or characteristics pertaining to the degree of confirmation of a posit, or to how it stands in the face of the empirical data that are considered relevant evidence.

According to *impartiality*, the cognitive values alone are constitutive of sound scientific judgments of theory acceptance.

Values (in addition to the cognitive values) can be regarded (on my account) as among the constitutive values of particular approaches to research, since they are built into the various objectives of approaches to research (O_i'), and they determine what counts as a potentially significant theory. In addition, commitment to values can properly lead to the expectation that assessments of the manifestation of cognitive values be submitted to tougher standards ("Rudner's argument": Chapter 4). Attributing these roles to values, however, does not cast doubt on the very idea of a cognitive value; rather it presumes that there is a proper role for cognitive values distinct from that of the other values. Longino adds to that characterization of a cognitive value that it is "a quality of [posits] ... that can serve independently of context as a universally applicable criterion of epistemic worth" (Longino 1996: 42).

This addition provides the thrust to her argument. Given it the idea of cognitive values indeed dissolves. First, what counts as a criterion of "epistemic worth" varies with the stance adopted towards a theory. The epistemic worth of a provisionally entertained theory (for example) is not subject to the criterion of empirical adequacy (though that of an accepted theory is), but it may be to that of fitting the constraints of a chosen strategy. Second, context is often important to interpret a quality like empirical adequacy, as illustrated in the preceding discussion of the empirical adequacy of theories developed under feminist strategies. Apart from the context of a strategy, to refer to a theory (of a domain) as empirically adequate is often hopelessly vague. Third, the ranking of the proposed cognitive values is contextually related to strategies, reflecting the way in which empirical adequacy is interpreted or the characteristics of the data selected for attention by the strategies. Full rather than wide-ranging explanatory power, the power to anticipate novel possibilities rather than predictive power, and capturing fine-grained details of relations rather that quantitative precision (for example) rank higher in the context of feminist strategies; *vice versa* in the context of materialist strategies. It is true, then, that the proposed cognitive values cannot function as criteria of "epistemic worth" independent of context (stance adopted towards a theory or strategy). Longino seems to suggest that this provides sufficient grounds for rejecting any clear distinction of roles between cognitive values and social, moral, etc. values, and that they should all be grouped together interacting with one another, as it were, at the same level.

My idea of cognitive values does not require the *kind* of universality present in Longino's addendum, as acknowledged in the preceding three points. It treats cognitive values as decisive criteria only for the acceptance of theories. They exercise their decisive role only after strategies have unfolded and prior stances have been taken towards theories. (Up to this point values and cognitive values intermingle without creating logical problems.) Cognitive values characterize the relationships that must exist between theory and data (and other theories) for a theory to be acceptable (to express some form of understanding of a domain of phenomena), and thus the criteria for choice among the various theoretical candidates consistent with the strategy in play. While choice of strategy reflects values, and that choice has implications concerning the interpretation and ranking of (some of) the cognitive values, once it is made, theories are accepted (when *impartiality* is respected) without further mediation of values. The idea of cognitive values needs only *this kind* of universality. I do not think that Longino's arguments undermine it.

Let us look at Longino's arguments or rather her questions, for her conclusions are presented as not yet settled. They raise important and novel considerations and involve two components. The first is to explore whether an alternative list of constitutive values, which she identifies, may be appropriate to adopt for a feminist approach to science, including for making judgments of accepting and rejecting theories. The second is to argue that Kuhn's list of cognitive values (accuracy, consistency, scope, simplicity, fertility) is no less grounded in (other) values than the proposed alternative list is, that only (social) values pick out one of the lists as that of "cognitive values." I will only discuss the first component here; criticism I would offer of the second brings in no new considerations.

A proposed alternative list of cognitive values

What makes a theory a good or acceptable one? Within a feminist approach to science, Longino sees the following criteria being deployed: empirical adequacy, novelty, ontological heterogeneity, complexity and mutuality of interaction, application to current human needs, and diffusion or decentralization of power (Longino 1995: 385; 1996: 45 ff.). Only the first item is a generally recognized cognitive value. She distinguishes the last two from the others as being "pragmatic." The others are treated as being on the same level, but with functions that are contextually variable. I think that the role of empirical adequacy needs to be sharply distinguished from the roles of the others, and that the other cognitive values must complement it.

Consider *novelty*, "models or theories that differ in significant ways from presently accepted theories, either by postulating different entities or processes, adopting different principles of explanation, incorporating alternative metaphors, or by attempting to describe and explain phenomena that have not previously been the subject of scientific investigation" (Longino 1995: 386). Novelty clearly is not a value pertaining to theory acceptance – the kind of warrant needed for acceptance is unavailable without extensive and prolonged research. It pertains at most to other stances taken towards theories, for example, provisional entertainment, or commitment to explore. Novelty itself is open to two interpretations. On the weak one it is linked with a desire to gain scientific understanding of hitherto neglected phenomena (Longino 1995: 387). Then what is important is not that the theory is novel, but that it is of these phenomena. The value in play here is a social one, "being of these hitherto neglected phenomena" and it may properly influence the strategy adopted in investigation. Novelty then is a trivial consequence of "hitherto neglected."

The strong interpretation refers to novel posits that are competitors with existing ones. Novel (original) competing theories are valued under certain conditions – for example, because one wants to test a current theory against more rigorous standards, or because currently available theories do not enable us to solve certain problems or do not fit strategies one wishes to adopt. Either way, the value of novelty is derivative; its value lies in being responsive to the considerations that underlie the desire for more rigorous testing (for example, "a suspicion of any frameworks developed in the exclusionary context of modern European and American science" – Longino 1996: 46), or in enabling hitherto unsolved problems to be dealt with, or in furthering research under one's favored strategies. Especially because value considerations can play a role (of which one may be unaware) in one's assessments of the acceptability of theories, it is often appropriate to raise the issues of higher standards of testing and of applicability of the theories to expanded domains of phenomena. A novel theory may contribute to cognitively significant considerations – not *qua* novel theory, but *qua* theory that over time develops in a certain way. Novelty is not a cognitive value. Considerations pertaining to it, however, may be essential to deploying the standards with which we assess the extent of the manifestation of the cognitive values. Neither is novelty *per se* a value that might be added to the list of constitutive values of an approach to investigation. A value on such a list, however, might point to the need for novel theories in certain areas, so that derivatively novelty might be an important value in making judgments of provisional entertainment of a theory and commitment to its development.

Consider *ontological heterogeneity*. A plausible argument was presented in the course of discussing Longino's criticism of LHM that the objective of a feminist approach to science (O_3) requires the development of theories that represent phenomena of human agency as ontologically heterogeneous. This requirement is reinforced when one attends to the more specific objective that, I pointed out, sometimes Longino considers: "to reveal gender"; for presumably the point of revealing gender is to guide projects aiming to overcome it, and thus to expand the possibilities for the exercise of the capacity of agency that are currently inhibited by gender. This argument stresses the importance of individual differences in investigations in the life and behavioral sciences, and the rejection of treating differences as departures from some type which is taken as representing something ontologically fundamental. She speaks of deriving from ontological heterogeneity the "rejection of theories of inferiority" (Longino 1996: 47). When she does so, she is not denying the "factual" claims of such theories on the ground of their discord with her adopted values, but pointing out that such theories have not been tested against theories developed under constraints dialectically linked with this value, so that, therefore, they have not been appraised under appropriately high standards (cf. the preceding discussion of LHM). It is thus appropriate to consider ontological heterogeneity as among the constraints on theories required by strategies that are adopted in view of choosing human agency and the expansion of the possibilities for its exercise as the object of inquiry – and thus to consider it as a constitutive value of the approach O_3. That theories meet this constraint derives from its mutually reinforcing links with the value of expanding human capacities for agency. Then, it remains for empirical inquiry to establish whether or not theories which manifest the cognitive values to a high degree can be developed under it. Similar conclusions holds for complexity and mutuality of interaction.

The other two criteria, the ones Longino calls "pragmatic" (Longino 1996: 48) require a different analysis. Consider *applicability to current human needs* – "directed towards meeting the human and social needs traditionally ministered to by women" (Longino 1996: 48). This can be taken as the recommendation not to apply, in practice, theories that do not inform attempts to redress current human needs, or as the recommendation to seek theories that can be applied to inform such ends. Then the value of addressing human needs functions as a criterion for the interest and significance of a theory, but not of its cognitive credentials (cf. the criticism of green revolution science in Chapter 8). It may also serve to focus our particular research interests *within* an approach to investigation defined by a certain object of inquiry, and perhaps occasion the identification of a new object of inquiry.

Diffusion of power is a property not of theories, but of the institutions and practices from which theories are produced. It sometimes has cognitive significance. In principle, the greater the range of perspectives brought to bear in the comparative assessment of theories, the higher the standards against which the manifestation of the cognitive values can be judged. Diffusion of power within research institutions and practices can be expected *ceteris paribus* both to bring a greater diversity of participants into the research process (likely to increase the range of perspectives brought to bear, but not necessarily under all conditions), and to give greater authority to hitherto marginalized perspectives (cf. the discussions of *autonomy*: Chapters 1, 4 and 10).

Thus diffusion of power refers to an institutional and practical value, deriving from its service to the rigorous assessment of theories in the light of the cognitive values, that is, to *impartiality*. One dimension of this is that of creating conditions in which contributions to knowledge from local, traditional and indigenous sources may not be ignored (Chapter 8). It also *ceteris paribus* serves the interest of applicability to current human needs.

Now consider *empirical adequacy*. As discussed on p. 212–14, its interpretation can vary with the strategies under which inquiry is conducted. Empirical adequacy is the value that a theory "fit" the relevant class of available empirical data, where the adopted strategy determines what counts as "relevant." Once a class of data has been picked out as uniquely relevant, it can bear evidentially only upon posits of certain types. But I have maintained that adequate testing of posits requires meeting standards that involve comparative assessment. This is especially pertinent when dealing with posits which – if considered incorporated into the stock of knowledge – challenge the presuppositions of moral projects (for example, feminism). Here empirical adequacy must function in concert with the other cognitive values. Consider T_M (theories developed under variants of the materialist strategies). They were criticized not for being empirically inadequate in the context of the materialist strategies, but for lacking (comparatively) explanatory, unificatory and anticipatory (predictive) power among the realm of phenomena that are described in the course of daily life and experience with action and intentional terms, and for lacking the resources to shape interpretive accounts of the successes, failures and bounds of application of competing theories. This criticism cannot be sustained unless empirical adequacy is complemented with several other cognitive values, so that Longino's own criticisms (as I have developed them) depend on commitment to something like a standard list of cognitive values.

Longino maintains that a cognitive value, such as empirical adequacy, does not have "a solely epistemic or cognitive basis." I concur: since adopting a strategy is partly rationalized in view of its mutually reinforcing interactions with certain social values, values contribute to some extent to the interpretation of empirical adequacy that one brings to bear on one's hypotheses. She goes on to say: "Empirical adequacy is valued for, among other things, its power when guiding inquiry to reveal both gender in the phenomena and gender bias in the accounting of them" (Longino 1996: 45). I take this to mean that empirical inquiry conducted under feminist strategies, *properly* responsive to the cognitive value of empirical adequacy (together with, I add, a few other cognitive values) has the power to reveal gender and, therefore, it is able to inform feminist projects, and that the inquiry gains value from that fact. The products of this approach to inquiry, but not generally those of O_1, can be adopted to inform practices of the feminist moral project. This gives them a value, lacking for the products of variants of the approach O_1, *provided that* they come to inform practices *effectively*. Effectiveness in application presupposes that the products manifest highly the cognitive values.

I mentioned on p. 212 that it is possible (but not likely) that, within a feminist approach to science, some theories expressing LHM might become soundly accepted. If they were, they would be no less valuable than theories that reveal gender, if soundly accepted, are. Their coming to be accepted would show that certain presuppositions of the feminist moral project could not be sustained. That would no doubt be a disappointment, but it would not be a ground to reject the outcomes of the inquiry, and to attempt to act contrary to them would be folly. The point, of course, generalizes: where feminist inquiry successfully reveals gender, to attempt to act contrary to the finding threatens to entrench what has been revealed. The role of the cognitive values concerns – once strategies have been adopted – only how the logical gap between data and posits is to be closed. This requires a clear distinction of role between cognitive values and other values.

I do not think that Longino's arguments undermine the distinction of roles between cognitive and social values (see Chapter 10). Equally, my arguments do not undermine, rather they reinforce, other conclusions she has reached: that scientific knowledge emerges from a process in which values are pervasive; that theories have desiderata linked with significance over and above cognitive ones; that values (since they influence the strategies adopted) and therefore social conditions partly (largely?) explain which theories come to be produced and accepted in the scientific community (Chapter 10); and that social values can be manifested in theories alongside the cognitive values (contributing to their significance for certain value complexes, but not their acceptability), since the applicability of some (kinds of) theories may serve particular value complexes especially well (Chapters 4 and 10). The crucial distinction of roles of the cognitive and (other) values can be reconciled with the pervasive role of values in the processes of science. It is not merely a formal reconciliation. The very intelligibility and rationale of gaining and attempting to apply systematic empirical understanding (science) depends on it.

The details of the reconciliation will be developed in the next chapter in which I attempt to articulate what can and cannot be defended of the three components of the idea that science is value free.

10 Science as value free: revised theses

I introduced theses of impartiality, neutrality and autonomy (in Chapter 4); but they concerned only inquiry conducted predominantly under the materialist strategies. It is now clear that this involves a serious oversimplification. At least in principle, and to some extent in practice, the objective of science – to gain understanding of phenomena – may (must?) be pursued with a number of different approaches, each deploying its own particular strategies. Where does this leave the view that the sciences are, or ought to be, value free? In view of the argument that no theories can become accepted outside of research conducted under strategies which are adopted in part because of their mutually reinforcing interactions with particular value complexes, does the idea of science as value free retain any sense? Or should it be dismissed simply as the allure of a way of thinking that disguises the role of values (the modern values of control) in the predominant approaches to modern science?

To address these questions, I will develop revised versions of *impartiality* and *neutrality* that leave open (in principle) a multiplicity of approaches to scientific inquiry. My answer will be *impartiality* can and ought to be sustained as a viable and important thesis; that *neutrality*, despite ambiguities that make it difficult to sustain in a clear-cut way, can be interpreted so as to be defensible; but that only fragments of *autonomy* can be sustained.

IMPARTIALITY

I proposed (Chapter 4) the following thesis of impartiality (in abbreviated form):

I'   1   The cognitive values are distinct from other values.

2   T is accepted of D if, and only if, T (of D) manifests the cognitive values highly in relation to E – where T and E fit the materialist strategies.

3   T is rejected of D, if and only if, T′ is accepted of D, where T and T′ are inconsistent.

Hence:

4   Values are not among the grounds for accepting and rejecting theories; and

5   Fitting the materialist strategies is the only antecedent require ment on T and E.

I′ builds in the materialist strategies; in it T and E represent theories and data (drawn from some domain of phenomena, D) that are constrained and selected respectively by the materialist strategies. Thus, by definition, judgments made under alternative strategies cannot be made in accordance with *impartiality*. This is unsatisfactory, for those who adopt alternative strategies aspire to making judgments of which (4) holds. They, too, do not want their judgments clouded by extraneous considerations; they want to separate the genuine from merely logical possibilities, to test their theories against the best materialist theories and to have their practical projects informed by sound empirical knowledge. We need a statement of *impartiality* that may apply to research conducted under any strategies.

I′ faces the further difficulty that (4) and (5) cannot be held together consistently, since (5) involves that the modern values of control are among the grounds for rejecting a whole class of theories, those that do not fit the materialist constraints, thus contradicting (4). I say "rejecting," rather than "not investigating," for a policy of not investigating is tantamount to rejecting. Then, only if the applicability of I′ were clearly restricted to theories of domains whose possibilities are exhausted by their material possibilities, could (2)/(3) be sustained as a value of scientific research practices. (I am assuming that (1) has been satisfactorily established.) Only if all the possibilities of phenomena were material possibilities, would it make sense to restrict the interpretation of *impartiality* in a way that presupposes the deployment of the materialist strategies. To put such a condition upon all scientific practice, prior to the outcome of inquiry over the long haul (without a sound argument for materialist metaphysics) would be an instance of "wishing" (Geertz: Chapter 9): treating the world, insofar as it comes into contact with our practices, as an object of possible control, as we may "wish" it to be, but which we have no sound ground to affirm that it is.

Could these problems be avoided by revising I′ so that acceptance of theories becomes relativized to strategies, as in the following (using the same abbreviated form as on p. 224–5 with I′):

I″   1   The cognitive values are distinct from other values, and they may be manifested in theories developed under a variety of different strategies.

2   T is accepted of D under S if, and only if, T (of D), in relation to E, manifests the cognitive values highly, and to a higher degree than any rival theory does – where T and E fit S.

3   T is rejected of D if, and only if, T′ is accepted of D, where T and T′ are inconsistent.

Hence:

4   Values are not among the grounds for accepting and rejecting theories.

Values may play a role in the choice of which strategy to adopt, but this is consistent with I″, which may be seen as a simple generalization of I′. Any theory accepted in accordance with I′ will also be accepted in accordance with I″ – except that now it will be accepted *under the materialist strategies*, and so accepted explicitly of domains of material possibilities of things. Theories, that do not fit the constraints of the materialist strategies, are now not rejected because they violate these constraints, but only rendered insignificant for not pertaining to the valued class of possibilities. Relativized acceptance need not lead to a problematic relativism of rational judgment. Different judgments reflect different foci of interest, different valued classes of possibilities, but that a theory is accepted under S (of D) depends only on its manifesting the cognitive values to a high degree in relation to the selected data – a judgment that can arguably be made (in principle) by everyone regardless of whether or not they actively conduct research under S and adopt the values linked with S. "T is accepted of D" is relative to S; but "T is accepted of D under S" is not relative to anything.

The relativizing of acceptance in I″ is consistent with a number of things that I have emphasized: the separation of the levels of strategies and theories, a role of values in the adoption of strategies which does not threaten that sound theory choices are responsive only to the data and the cognitive values, the claim that there is no way to pursue the general objective of science, to gain understanding of phenomena (O) – especially its component: to encapsulate the possibilities of things and the ways to realize some of them – except under the strategies of a particular approach (O$_i$: Chapter 5). On the other hand, it is marred by a serious ambiguity. Does "rival theory" designate a theory that fits the constraints of S, or one developed under any strategies? If the former, then "accepting T of (D)" would be, in effect, "accepting that T (of D) is the best theory (of D) developed under S"; and "rejecting T"; would be "accepting a better theory (inconsistent with T) developed under S". This would avoid rejecting rivals because of their lack of fit with S and thus in part for reasons that include values. If the latter, there would be no need to resort to relativizing acceptance and rejection of theories. If we can compare the degrees of manifestation of the cognitive values in theories developed under different strategies, why would we be content to compare them only in theories developed under the same strategies? Why keep the theoretical products of different approaches separate, without critical interaction with one another, as if they are not all framed by O? There is a reason, but it does not provide the whole story. Unless a strategy unfolds of its own impetus, it is not likely to develop theories that manifest the cognitive values to a high degree; fruitful comparison of theories across strategies requires sufficiently developed theories (in relevant ways of the same domains). At least sometimes, there are two moments to the acceptance of theories: acceptance that T is the best theory of D under S, *and then* unqualified acceptance of T (of D).

Sometimes we do compare the cognitive merits of theories developed under different strategies, and judge theories developed under one strategy to manifest the cognitive values of comparable domains of phenomena more highly than those developed under another (incompatible) strategy.[1] When we do this the cognitive values "encapsulation of the possibilities of a domain of phenomena," "explanatory power" and "capability to define the limits of application of a theory" are likely to be most salient. Since a particular approach is linked (in mutually reinforcing interactions) with particular values, the manifestation of "encapsulate the possibilities ... " *prima facie* may be expected to be restricted. Attention to it alerts us not to identify readily the possibilities afforded by the phenomena of a domain with those possibilities of the phenomena that are of interest in light of these values; to attempt to define the limits of applicability of theories developed under a particular strategy; and to raise the question of whether the lexicon of the strategy being followed has the resources to characterize all the humanly and environmentally relevant effects of application of theories developed under the strategy (Chapter 8, and the later sections of Chapter 7).

It is always pertinent, in principle, to ask: concerning specified domains of phenomena that we are investigating, could we adopt alternative strategies that would enable us to gain theories that would manifest the cognitive values (especially the three just listed) more highly? If the answer is "yes," then *ceteris paribus* that is sufficient reason to adopt the alternative strategies for the investigation of those phenomena, regardless of the (social) values linked with the initial strategy, and it might also provide the occasion to reassess commitment to these values. In this way the cognitive values may also play a role in considerations of the choice of strategies for investigating certain domains of phenomena, but that does not annul the mutually reinforcing interaction of the strategies with a set of values. Links with certain values and not others, thus, may be conducive for gaining theories that manifest the cognitive values most highly of specific domains – as illustrated in the ways that the modern values of control are conducive for gaining theories that encapsulate the material possibilities of things and that represent fully the possibilities of certain domains in which relevant human influence is absent, feminist values for gaining theories about human cognitive abilities (Chapter 9), and the values of authentic development for gaining access to possibilities (in impoverished countries) of agricultural practices that would serve both social justice and environmental integrity (Chapter 8).

We need a formulation of *impartiality* that, unlike I', does not build in the adoption of a particular strategy and that, unlike I'', permits impartial judgments in theory choices made across strategies. To this end we might try replacing I''(2) (again maintaining the abbreviated form) with (2a) and making the obvious related adjustment to (3).

> (2_a) T is accepted of D if and only if T, in relation to E, manifests the cognitive values highly – where T is constrained by S, and some of its rivals are constrained by S_1, S_2, ... ($S_i \neq S$), and E includes the data selected under all of S, S_1, S_2 ... ; and making the obvious related adjustment to (3).

But E cannot include all these data. Data are sought out and recorded under a strategy, from which the lexicon of observational reports is obtained. Generally, competing strategies are incompatible and (to some extent) their lexicons are incommensurable. Recall that where an Aristotelian scientist (A) observes an instance of inhibited free fall, a Galilean (G) observes a damped pendulum (Chapter 7). G's theories do not have to "fit" A's data, and *vice versa*. In G's theories there are no inhibited free falls, and what A describes as an inhibited free fall is considered to be in fact a damped pendulum. A theory must – so to speak – only fit data that express facts of phenomena of D. Methodologically this means that, while a strategy remains actively in contention, theories developed under it are required only to fit data selected under it. This raises questions about D, the domain of phenomena of which theories are in competition. We grasp phenomena in scientific 'worlds' and/or social "worlds" (Chapter 7), with lexicons developed either in scientific practices or the social practices of daily life and experience. If the items of D are characterized solely with scientific lexicons, then generally the question of comparing theories across strategies will not arise, for a phenomenon characterized with the lexicon of S_1 will be part of a different 'world' from that of one characterized with the lexicon of S_2; theories developed under the different strategies will not compete of domains in the same 'world.'

There can be competition only when there is a shared characterization of the items of D – in a lexicon of social practices of daily life and experience. G and A competed of domains that included such phenomena as children's swings, flights of arrows and planetary movements. Each proposed applications to phenomena of this social "world," of which both their scientific 'worlds' were "sub-worlds." The categories of this "world" include those of both 'worlds'; indeed scientific activity in general is unintelligible when abstracted from the "world" in which it gains the necessary conditions to be pursued, and the contexts in which to be applied and in which its theories gain significance. Thus, with these categories it becomes possible to formulate that a phenomenon, f, characterized with the categories of a common social practice (for example, playing on a swing) is identical to f, characterized with the categories of A, or to f, characterized with those of G. Let D, a domain of the "world," be represented by $\{f_1, f_2, \dots\}$. When A is applied to D, it represents it as $D = \{f_1, f_2, \dots\}$; and G as $\mathbf{D} = \{\mathbf{f_1}, \mathbf{f_2}, \dots\}$. Let E be the class of data formulated with the lexicon used in D; and $\mathbf{E}$ with the one used in $\mathbf{D}$. The items of E and $\mathbf{E}$ are largely (not entirely) incompatible. G is accepted of D because it manifests the cognitive values, in relation to $\mathbf{E}$, more highly than A does in relation to the items of E (Chapter 7). It is accepted of D because it is accepted of $\mathbf{D}$. This judgment does not require that either theory be expected to fit data other than those selected in accordance with the strategy under which it developed.[2] Generally the question of comparison of theories across strategies arises only in the context of the application of theories to domains of phenomena in the realm of daily life and experience. It is often of great importance when facing practical (technological) applications of theories, especially questions about their legitimation.[3]

I now offer the following statement of *impartiality*:

I 1   The cognitive values are distinct and distinguishable from other values, and they may be manifested in theories developed under a variety of different strategies.

2   T is accepted of D under S if, and only if, T is accepted of D under a strategy S; and so, in relation to E, manifests the cognitive values highly according to the most rigorous available standards; and to a higher degree than any rival theory manifests them in relation to the data appropriate in the light of the strategy under which it developed – where T meets the constraints of and the items of E have been selected in accordance with S, and some of the rivals are (were) developed and appraised under different strategies.

3   T is rejected of D if, and only if a rival theory (T′) is accepted of D, and T and T′ are inconsistent, regardless of the strategies under which T′ developed.

Hence:

4   Values and assessments of a theory's significance are not among the grounds for accepting and rejecting theories.

I articulates nicely the value of impartiality.[4] It is well expressed in the practices of systematic empirical (scientific) inquiry to the extent that theories accepted in them are accepted in accord with *I*, and that they are conducted so as to increase the number, proportion and variety of theories accepted in accord with *I*. Adopting *I* requires attentiveness to the conditions under which its expression can be furthered. It is also a value of scientific practices, I propose, that all theories that inform a practical (technological) application and the (factual) presuppositions of the legitimacy of the applications be accepted in accordance with *I*.[5] Judgments made in accord with *I* will be preceded by judgments made in accordance with I″. In the case that D is not constituted by phenomena in the realm of daily life and experience, *I* and I″ (or I′) tend to collapse together, although if D can be maintained to be open in principle to investigation under a different strategy, accord with *I* will not be reached prior to comparative testing against theories developed under the different strategies. Then, before a proper comparison can be made, theories under the respective strategies must be appropriately developed.[6]

Impartiality is a demanding thesis and its manifestations are not casually upheld. There are, however, many theories (of specified domains) that have been accepted in accordance with *I*, and that also are significant for many value complexes. Indeed, exemplary cases have been developed and consolidated under the materialist strategies (though *I* is not always a "fact" of theories actually accepted under these strategies). *Within many spaces,* whose kinds of boundary conditions we can broadly specify, phenomena have been soundly grasped – as evaluated by the cognitive values – in terms of being generated from underlying law, process and structure, and their possibilities are exhausted by their material possibilities. The world *is* this way; we have not taken it to be this way because of our desire to control it. In the long run, *part* of the reason to continue following the materialist strategies is the considerable and on-going empirical success gained from following them, the success of coming to accept theories (of certain domains) in accordance with *I*.

Separate roles for cognitive and other values

When one adopts the materialist or any other strategies, in effect one lays out in the most general terms the kinds of phenomena and possibilities to be investigated. There is nothing *logically* improper about social values strongly influencing this choice. *Then,* the acceptability of theories constructed under the strategies is judged in the light of the data and the cognitive values (where having been developed under a particular strategy is not a cognitive value: Chapter 5). The important thing is to keep the roles of the social and the cognitive values separate. Their different roles reflect different (logical) moments connected with making theory choices. At *one* moment, when we ask: "What characteristics must theories have to be provisionally considered?", strategies play the key role. They serve to eliminate theories that do not fit their constraints. They function (logically, not necessarily temporally) first. In principle, with respect to a given domain of phenomena, an array of incompatible theories will fit the constraints; the play of the strategies is insufficient to determine which theory to accept. Then (logically) at the *second* moment, when we ask: "Which (if any) of the theories, with these characteristics, is to be accepted?", one of the theories from the array may be accepted. Here, the play of the cognitive values, in the light of the empirical data and other accepted theories that are available, is (according to *I*) decisive.[7]

At the moment of concrete theory choice only the cognitive values properly play a role. If, given the data available in the current state of investigation, the cognitive values do not suffice to make a clear choice, no theory may be considered soundly accepted and the matter must remain open to further investigation. If, in actual fact, a theory is accepted under these conditions, then a value has played a role (improperly) *alongside* the cognitive values in making the judgment of acceptance. This does not preclude values playing a proper role, however, not *alongside* the cognitive values, but at the moment when the strategies function, interacting dialectically with them. Strategies are adopted because of interest, typically derived from values, in the possibilities that may be encapsulated in theories constructed and consolidated under them. Whereas strategies lay out the general features of the possibilities of interest, a properly accepted theory encapsulates what the genuine possibilities are.

One may adopt strategies, then, because of their relationship to values, because they may enable us to encapsulate possibilities that may inform one's moral and social projects. That does not imply that one eliminates from consideration theories that do not fit the strategies because one believes that they are false; rather one may do so because they do not provide a means to identify the possibilities of interest. One tends to adopt strategies under which it can be expected that significant theories will be developed. Now, a significant theory is also a soundly accepted one; but a theory, that is soundly accepted of certain domains of phenomena and that reliably encapsulates their possibilities, may have no relevance for one's practical projects. (Dismissing a soundly accepted theory as insignificant does not imply contradicting that theory.) Adopting a particular strategy does not (and cannot) commit one to accept any theory; rather it frames the quest to construct and consolidate theories of certain kinds, but it provides no guarantee that the quest will be successful. Not all strategies are like the materialist strategy, such that the world lends itself to revealing certain of its possibilities (many of which can be realized through our interactions with it) in the course of research under them. Persistent failure to develop under given strategies theories, which manifest the cognitive values to a high degree, is *ceteris paribus* a decisive ground to abandon those strategies. Thus, the adoption of strategies is not only linked dialectically with values, but also is under long-term empirical constraint.

A theory is properly accepted (rationally believed to encapsulate the possibilities) of a domain only if it manifests the cognitive values to a high degree according to the highest standards for assessing the degree of their manifestation in theories. The values, that make these possibilities interesting and that may motivate the provisional entertainment of theories that fit the strategies that are dialectically linked with them, play no (proper) role in judgments of acceptance.[8] They do (ought) not function alongside the cognitive values. The distinction and separation of moments is methodologically and logically essential. The answer to "On what grounds is T (of D) accepted?" properly includes no appeal to values, but the answer to "Why does the scientific community investigate theories that are in line with certain strategies?" will typically do so; appropriately so. Although with the passage of time that will not be sufficient by itself, and will need to be bolstered with evidence that following the strategies leads to success in consolidating theories that manifest the cognitive values to a high degree.

The answer to "How did it come about that T (of D) became accepted?" will also involve reference to values – those which sustained the research activity and motivated following the strategies. Values play a causal role in the development and consolidation of theories. That is consistent with I, for I is a thesis about the sound acceptance of theories, not one about how to explain that a theory came to be accepted. Pointing to the grounds for soundly accepting T (of D) is not the same thing as explaining why T came to be soundly accepted. The latter requires also reference to the process of investigation: Why, for example, was T provisionally entertained in the first place? Why did certain investigators commit themselves to its development? Why was D considered of interest? Theories which satisfy I are the residue of a value-implicated process. Thus explaining that T has come to be accepted will involve reference to values, but this is irrelevant to whether T is accepted in accordance with I or not. The question: "Why is T accepted?" is ambiguous between "On what grounds is T accepted?" and "How has it come about that T is accepted?" Some social constructionists have assumed that good answers to the latter annul or need not pay heed to answers to the former, and thus have erroneously concluded that the role of values (and the power that may support their role) in good answers to the latter shows that *impartiality* cannot be realized.

Regardless of what values pertain to the causal history in which T becomes accepted, I may be satisfied, but that causal history may also bring it about that T be an object of value only within certain value complexes, especially those that interact in mutually reinforcing ways with the strategies under which T developed. Accepting T does not imply its general significance across value complexes. At the same time attending to the causal history within which it becomes accepted is not irrelevant to judgments about whether T has been accepted in accordance with I, for that history can alert us to whether T has been tested against an appropriate array of competitors, including (in principle) competitors developed under competing strategies. More generally, it provides the evidence pertinent to judging if the degree of manifestation of the cognitive values in T has been estimated according to the most rigorous available standards. Failing to attend to the causal history may hide the covert role that values may actually be playing alongside the cognitive values when appropriate competitors are not entertained.[9] In many of these situations, a mistaken inference is covertly made from accordance (discordance) with I' to accordance (discordance) with I.

The judgment that T (of D) is accepted in accordance with *I* is logically independent of the values that are held. Thus, when well made, it is binding generally regardless of the values held, regardless of the possibility that the adoption of certain values may be a necessary condition for engaging in the research activities (or even for having the skills and capabilities to do so) that enable such a judgment to be made, and regardless of the possibility that the theory may be significant only where these values are adopted. Holding different values, then, is not a ground for denying the cognitive value of such a judgment, or for acting on the basis of beliefs that are inconsistent with it. But holding them may void the judgment of the significance of the theory, and provide a reason to engage in investigation exploring different classes of possibilities by adopting different strategies.

Rejection of theories

So far in this section I have focused on the acceptance of theories. When we turn to their rejection additional matters arise. If T has been soundly accepted of D, then, in accordance with *I*, theories that are inconsistent with T will properly be rejected. In such cases, rejection follows from the data and deployment of the cognitive values. Adopting strategies involves another kind of "rejection." Not only does it underlie choosing those (embryonic) theories that are to be provisionally entertained and adhered to for the sake of their development, but also characterizing those that are simply not to be considered, those that are to be "rejected" out of hand or "side-lined." What is the status of the side-lining of theories when, for example, under the materialist strategy theories that deploy teleological, intentional and sensory categories are side-lined?

Several cases need to be distinguished. In the first, the side-lining reflects simply a lack of interest in the domains of phenomena and classes of possibilities that such theories might illuminate, and thus the judgment that these theories will be insignificant. *Per se* it has no cognitive repercussions. It may come to have some. Suppose that the side-lining of a type of theories is done by a group of scientists who adopt S, but that in the scientific community as a whole other groups adopt different strategies, so that we would expect that the appropriate array of theories would be developed to test the accord with *I* of T's eventual acceptance. Then, side-lining would be by a group for the sake of the serious unfolding of S, and it would be accompanied by the active tolerance of groups adopting different strategies, and leading towards an eventual critical interaction with the theories produced under the other strategies. If the whole scientific community adopts S, then side-lining amounts to rejection that does not accord with *I*.

In the second case, side-lining does have a clear cognitive dimension and is equivalent to rejection. In *I*, a theory is said to be accepted of a particular domain, or domains, of phenomena. A theory, accepted under the materialist strategies in accordance with *I*, will be accepted of a wide range of such domains. It is reasonable to affirm that the success of science conducted under the materialist strategies has amply demonstrated that, concerning spaces with certain kinds of boundary conditions, phenomena can be well grasped under these strategies, legitimating the rejection of theories that do not fit the constraints of the strategies. This is a kind of meta-induction. It supports following the materialist strategies exclusively when dealing with phenomena in such spaces, and they may be the vast majority of phenomena that gain the attention of physicists in the normal course of events. I do not know how to define precisely "certain kinds of boundary conditions,"[10] but the meta-induction definitely does not extend to spaces in which human agency is a relevant causal factor, for there the superiority of materialist to intentional explanations has not been demonstrated (Chapters 6 and 9).

The third case concerns spaces where human agency is a relevant causal factor. Of course, one may have no interest in investigating phenomena in these spaces, so that then side-lining theories that deploy non-materialist categories amounts to simply opting out of research on them. If one does investigate them, however, there is no cognitive ground to reject non-materialist theories out of hand. The only grounds would be rooted in values. In Chapter 1, I discussed the long tradition in which science has been identified with that mode of empirical inquiry that follows the materialist strategies. If, in the name of doing "science," the scientific community attempts to understand the phenomena of agency through following the materialist strategies exclusively, then *de facto* values will be among the grounds for the theories that come to be accepted in the community, specifically those values, commitment to which partly led to the out-of-hand rejection of non-materialist alternatives, and thus to the failure to test one's theories against such alternatives.[11] Then, any degree to which the cognitive values become manifested in these theories will not be measured against the highest available standards – an ironic twist to the tradition that insists on science as value free.

The term "modern science," I pointed out (Chapter 5), typically connotes both systematic empirical inquiry and inquiry conducted under the materialist strategies. The idea of science as value free has been articulated as part of the self-understanding of the tradition of modern science, and both connotations permeated the sources of the idea. And it is true that, under the materialist strategies, we have had great success in coming to accept, in accordance with *I*, theories of a wide range of spaces. Nevertheless, for an important range of spaces – where human agency is a relevant causal factor – inquiry, conducted exclusively under the materialist strategies, cannot produce theories that are accepted in accordance with *I*. To be accepted in this way the theories of the phenomena in these spaces must be tested in comparison with theories (posits) developed under other strategies, and some posits (for example, "No lost possibilities": Chapter 8) are not amenable to investigation under the materialist strategies. Where inquiry is limited to the compass of the materialist strategies posits like these, that may function in the legitimation of applications of theories, function as if a priori, even though they are open to empirical inquiry (Lacey 1997c). *I* is a value of the practices of systematic empirical inquiry. That is why I have stipulated that "science" be identified with "systematic empirical inquiry" and not with "modern science" with its dual connotations. I push my stipulation because *impartiality*, an ideal carried by the tradition of "modern science," can be realized generally only within the widened notion of "science," which admits of systematic empirical inquiry under a variety of strategies, among which the materialist strategies perhaps have pride of place.

NEUTRALITY

As I have emphasized, it is important to keep the roles of the social and the cognitive values separate, and that of the social values confined to its proper moment. Nevertheless, if social values have a role at all then it follows that theories, even when accepted in accordance with *I*, may not be *neutral*, at least in the sense of "evenhandedness in application," of being significant for any viable value complex, N' (3) (Chapter 4). They may not be generally significant because the possibilities encapsulated in a theory may be of little interest for practices which express values other than those linked with the strategies that generated the theory, because any practical application of the theories may require conditions and have consequences that undermine the embodiment of the preferred values; as the possibilities of increasing crop yields that also involve the seed becoming a commodity have little interest where prevailing social values emphasize enhancing local well being, agency and community (Chapter 8).

I introduced provisionally (Chapter 4) the following thesis of *neutrality* (in abbreviated form):

N′   1   Accepting T in accordance with I′ implies no value commitments.
     2   Accepting T neither undermines nor supports holding any one of
         the range of viable value complexes; and
     3   If T is accepted in accordance with I′, it is significant to some
         extent for any viable complex.

The discussion of the seed (Chapter 8) provides an example of the breakdown of N′(3). It shows that theories, accepted in accordance with I′, need not be significant (even in principle) across all viable value complexes. Can N′ be revised to meet this counter-example? The discussion of the seed also points to the need for a statement of *neutrality* that applies where *I*, rather than just I′, is applicable. In addition it provides a hint worth exploring about how *neutrality* might be revised, when it points out that the same genetic theory that informs the research on hybrid seeds that are the basis of the green revolution is also compatible with comparable increased crop yields being obtained from appropriately selected pure varieties of seeds.

Perhaps *neutrality* should be restricted to "fundamental" theories – and then affirm that fundamental theories apply equally both for value complexes that highlight the modern values of control and for those in continuity with local traditions. Fundamental genetic theory , for example, does apply for both, but is it equally significant? It seems clear that it is not, since it is more readily technologically applied, and the material and social conditions required for its development are more readily available (without undermining the value complex adhered to), where the modern values of control are adhered to. Significance is a matter of degree, and this must be built into any defensible version of *neutrality*. I will do this, but I resist formulating a thesis of neutrality pertaining to "fundamental" theories since I do not recognize a sharp separation of "fundamental" and "applied" science.

Instead, I suggest substituting in place of N′(3) the following, which does not require all theories accepted under the materialist strategies to be significant for all viable value complexes, but only that for any viable value complex there are some accepted theories that are significant for it:

    3′   For any viable value complex, there are, in principle, some theories
         accepted in accord with I′ that are significant *to some extent*; and
         (perhaps) the more wide-ranging ("fundamental?") a theory, the
         more it is significant to some extent across a greater range of vi-
         able value complexes.

Replacing (3) by (3′) meets the objection that was raised but, like (3), it does not satisfy those (see Chapter 4) who think that N′ does not capture a strong enough sense of "evenhandedness." It is compatible with vast differences in the extent of significance for different value complexes (and even, for some, when all things are considered an overall negative value). The significance of theories accepted, in accord with I′ or I, under the materialist strategies is assuredly greater in those value complexes that contain the modern values of control (Chapter 6). The mutually reinforcing interaction between the materialist strategies and the modern values of control sets a bound to the scope of the "evenhandedness," or comparable significance, of these theories. I remain unable to move towards a stronger reflection of theories having comparable significance without falling into paradox (Chapter 4). (3′) recognizes that the significance of theories can transcend the context of their origin and development, and (with its addendum) it acknowledges that there is something about properly conducted "fundamental" science that in general carries a measure of *neutrality* with it. That is not a trivial matter. It also reflects properly that it is difficult to generalize about *neutrality*: methodological considerations and judgments made in choosing theories (including that a theory is accepted in accordance with *I*) alone cannot settle what is the range of value complexes across which a theory has significance and to what extent. Empirical investigation, involving social factors, is needed case by case.

Modifying N′ so that (3) is replaced by (3′) changes the focus of *neutrality* – its key third component now points to a (desired) feature of the research practices engaged in under the materialist strategies rather than to a feature of each accepted theory itself. In order to make this explicit, I revise N′ as follows:[12]

N″ Scientific practices conducted under the materialist strategies generate theories (accepted in accordance with *I*) such that:
  1 accepting these theories implies no value commitments;
  2 accepting them neither undermines nor supports holding any one of the range of viable value complexes; and
  3′ for any viable value complex, there are (in principle) some accepted theories that are significant to some extent.

Given that in all cultures some practices of control are valued (however subordinated they may be to other social values, Chapter 6) and that numerous phenomena (regarding which relevant human causal influence is lacking, Chapter 6) are shared in the daily life and experience of all cultures, we would expect that N″(3′) would be highly manifested – and it is; so also is it that only where the modern values of control are part of a value complex are theories accepted under the materialist strategies routinely significant to a high degree.

Evenhandedness

Shifting the focus from individual theories to scientific practices is fruitful. Part of the appeal of the initial idea of *neutrality* is that science can serve, more or less evenhandedly, the interests of all (viable) value complexes. The shift keeps that idea alive. But, if scientific practices are limited to those conducted under the materialist strategies, the "evenhandedness" expressed in N'' does not amount to comparable significance across viable value complexes, for overwhelmingly theories are of greater significance for those complexes that include the modern values of control. Can we identify such a sense of "evenhandedness" if we do not limit scientific practices in this way? And, in doing so, can we also keep alive the hope that where the interests of value complexes clash, aspects of the conflict that are implicated in posits about the world – about nature and human nature, about what is and is not, about what is and is not possible – can in principle be settled decisively (at least in the long run) through investigation that is systematic and empirical, and aiming to produce theories (posits) that accord with *impartiality*? We find hints again in the discussion of "no lost possibilities" (Chapter 8).

Recapitulating: in view of its adverse social and ecological effects, the significance of the scientific theories that inform the green revolution is largely restricted to the interests of the market and of other values closely linked with the modern values of control – and does not extend, for example, to those viable value complexes that highlight local well-being, agency and community. It was objected that these latter value complexes are not viable because their presuppositions contradict the posit: "Outside of green revolution (and successor) technologies there are no possible mechanisms to increase crop yields commensurate with the need to meet the basic food needs of the poor peoples of the world – hence adopting these technologies, and implementing the necessary social reorganizations, involves no lost possibilities relevant to addressing the need to provide food." Clearly, if "no lost possibilities" were a posit accepted in accordance with *I*, any value complex whose presuppositions contradicted it would not be viable. But research, conducted entirely under the materialist strategies, could not establish it as such a posit, for that research does not investigate developments that might be made under different strategies, for example, strategies that involve a dialectical interplay of traditional knowledge and research under the materialist strategies.

It is difficult to engage in empirical investigation of "no lost possibilities" because it necessitates the comparison of results gained from incompatible strategies, because relevant alternative strategies are not developed to anything like the extent to which the materialist strategies are, and because the conflicting strategies are respectively linked in mutually reinforcing relations with social values and socio-economic projects with little contemporary access to power (Lacey 1997c). Thus, gaining the material and social conditions for conducting the relevant research may be virtually impossible because power denies their provision. Put simply, the material and social conditions required for the relevant research are incompatible with the interests of the market. Their absence, thus, may reflect not the inherent inability of alternative forms of research to develop but the hegemony of the power of the market and its related forces. Whatever the explanation for the actual underdevelopment of research under alternative strategies may be, without it "no lost possibilities" cannot be accepted in accordance with *I*; so that where this posit is accepted with virtual certitude one rightly suspects that it is accepted because it is a presupposition of value complexes containing the modern values of control, or that it is functioning as a disguise for the "realities" of power. Without it, furthermore, research cannot be expected to generate theories of much general significance for those who adhere to value complexes in which the value of control is subordinate.[13]

Part of the appeal of the initial idea of neutrality is that judgments about posits like "no lost possibilities," which are at the heart of social controversies, could be grounded in the outcomes of systematic empirical inquiry – and that in this way "evenhandedness" ("comparable significance") would be obtained across the widest range of viable value complexes. It is now clear that research conducted solely under the materialist strategies cannot adequately reflect "evenhandedness," and take us to a thesis of neutrality more robust than N''. Perhaps it can be reflected more robustly, while not rejecting what is sound in N'' when scientific (systematic empirical) inquiry is taken (in principle) to include practices conducted under a variety of strategies.

New statement of neutrality

In the light of these considerations, I offer the following formulation of neutrality:[14]

N Practices of scientific (systematic empirical) inquiry conducted under a variety of strategies generate theories, that are accepted in accord with *I*, such that:

1   accepting these theories implies no value commitments;

2   accepting them neither undermines nor supports holding any one of the range of viable value complexes; and

3   in principle, for any value complex that remains viable as the stock of theories (accepted in accordance with I) expands in the course of research that puts its presuppositions to empirical test:

(a) there are some accepted theories, developed under materialist strategies, that are significant to some extent; and

(b) there are some accepted theories, some of which may be developed under non-materialist strategies, that are highly significant.

N incorporates what is sound in N″, and it does – recognizing that control has some place in all value complexes – grant a unique kind of recognition to the materialist strategies in (3a). It keeps contact with the initial idea of neutrality, including, for example, by retaining item (1), which it effectively shares with N″. But N also goes beyond the initial idea and removes any paradox that arises from (covertly) interpreting it in a way that treats particular value-linked strategies (materialist) as the source of neutrality. It also introduces a measure of interpretive complexity; *neutrality* gains higher manifestation in the context of inquiries conducted under multiple strategies, enough strategies to enable the presuppositions of currently viable and adhered to value complexes to be submitted (in some measure) to empirical investigation. The gain is a coherent idea and a value that is worth aspiring to manifest more fully, even though it may have little current manifestation and its prospects for fuller short-term manifestation may be dim in the light of current economic "realities" (Lacey 1997c).

I have maintained that human flourishing, in all of its dimensions and variations, and for as many people as possible, is my fundamental point of reference (cf. Kitcher 1993: 391; Dupré 1993: 244, 264). How adequately this value can coexist in the same value complexes as the modern values of control is a matter open to on-going investigation and controversy about the appropriate forms of empirical inquiry in which to conduct it. There is not a unique value complex to which adherence is required for living a flourishing life; equally not any fancied and articulated value complex can nourish flourishing, but only those whose presuppositions fit with what is permitted by nature and human nature. I am trying to explicate an idea that scientific (systematic empirical) practices be neutral across all (actually adhered to) value complexes whose presuppositions fit with what is permitted by nature and human nature. The best criterion we have of this is that the presuppositions be consistent with what has been accepted in accordance with I – hence the role of viable value complexes in N.

As I have defined it, "viable" is an historical variable: because the realm of genuine human possibilities changes with historically variable social structures, and because further investigation of the presuppositions of a currently viable value complex may render it non-viable. Thus, if for a particular (currently) viable value complex there are no accepted theories that are highly significant, this may reflect only that subsequent research would render that complex non-viable – though obviously this could not be known prior to conducting the research It is hardly an undesirable limitation of the scope of *neutrality* that it not extend across those value complexes whose current viability derives only from the failure to carry out the research which, if conducted, would produce posits accepted in accordance with I that are inconsistent with their presuppositions. On the other hand, it is undesirable that it not extend across value complexes that are viable and currently adhered to by some groups. However, where the values of the complex are not highly manifested because of the play of power, the general idea of neutrality resists the identification of the "viable" with what the "realities" of power make the "socially (historically) sustainable" (Chapter 4) or with the dominant tendencies of the moment. *Neutrality* should extend across those value complexes that are not only viable and currently adhered to by some groups, but also remain (or would remain) viable as the stock of theories (posits) accepted in accordance with I increases (or would increase) in response to research aiming to test the presuppositions of the complexes empirically and thence to assess the consistency of the presuppositions with the expanded stock of theories accepted in accordance with I – hence the long qualification introduced in $N(3)$.

N thus would become more highly manifested if research were to be conducted so as to expand the stock of theories accepted in accordance with I in a way that submits the presuppositions of all currently adhered to value complexes to empirical testing. Clearly any theories (posits) that become accepted in the course of research that tests, and comes to support, the continued viability of a particular value complex are likely to be significant for the complex. As illustrated with "no lost possibilities," such research needs to be conducted under a multiplicity of strategies. For N to become highly manifested enough strategies would have to be in play to provide an opportunity for the adherents of all (many) viable value complexes to attempt to consolidate significant theories and for their presuppositions to be put to some measure of empirical test. Conducting the relevant research requires material and social conditions, which may only be available where specific social values are highly manifested, such as the value of nurturing a wide variety of ways of leading flourishing lives that are open to as many as possible. Those conditions may not be readily available where the modern values of control are well entrenched. Then *N* would not be endorsed as a value, though N'' might be. Speaking loosely, *N* itself is not "neutral" (cf. Harding 1998). If scientific practices are valued in general across value complexes, it is not *because N* represents a value in all of them.

Certainly *N* does not represent a "fact" of current scientific practice, as illustrated in the case (and cases of this kind can readily be multiplied) of the science that informs the green revolution. Where scientific practices are confined to those carried out under the materialist strategies, *N* cannot be highly manifested, for their products are especially significant for value complexes that contain the modern values of control. Given current trajectories it would be rash to anticipate the fuller manifestation of *N*. Indeed, it does not represent a value endorsed in much contemporary scientific practice, although I think that arguments, for example, that feminist perspectives serve not to bring values illicitly into science, but to detect and challenge "bias" already present in scientific practices and products, indicate endorsement of something like *N*, as well as of *I* (Chapter 9).

Neutrality: *the tension between N and* N''

Perhaps it is N'' that articulates best the value of neutrality widely endorsed in the tradition of modern science – or perhaps only its first two items? More likely, I think, *neutrality* is actually taken to be manifested sufficiently (by its proponents) as long as more than one socially sustainable value complex (with endless individual variants), for which scientific theories are highly significant, is in play in the political arena, where each of them contains the modern values of control; and collectively their scope of manifestation continues to expand to include more and more peoples from "underdeveloped" societies. *Neutrality*, so understood, is framed by "progress" or "development", by the modern values of control. The essential link with these values is not taken as a sign of the absence of *neutrality* (as such a link with any other values would be) only because it is assumed that the modern values of control have become universal values. With the unfolding of modernity, numerous pre-modern value complexes have been rendered historically unsustainable. Some have been shown to be non-viable; others have simply succumbed to superior power or the effects of invasion. It does not violate the claim of "evenhandedness" that scientific developments have revealed some value complexes to be non-viable. It does violate the claim that scientific theories have served as instruments to inform practices of the powerful that have destroyed the conditions for sustainability of some viable value complexes, especially when – as in "no lost possibilities" – scientific practices provide no investigations to counter ideological claims affirming the non-viability of a value complex.

N″ is too weak to sustain "evenhandedness" ("comparable signifi-cance"). One may argue against the value of "evenhandedness," maintaining that the value of scientific inquiry derives, for example, from the links with the modern values of control or from individualist accounts of human nature. I do not rule this out of hand. My point here is that such links with social values should not be disguised under the cloak of *neutrality* and that, where the argument draws upon the claim that the modern values of control *ceteris paribus* lead to enhancements of human flourishing, it is based on a posit that has not been accepted in accordance with *I* (and could not be as long as the conduct of research is confined to the materialist strategies).[15] If one concludes that *N* does not represent a value desirably expressed in modern scientific practices and that the *neutrality* articulated in the tradition of modern science just boils down to N″, then it would follow that modern science is not seriously "evenhanded." That, however, would carry a sting only among those who seriously entertain value complexes in tension with the modern values of control.

While I do not doubt that most members of the scientific community would endorse an articulation like N″ rather than *N*, I suspect that many of them would be reluctant to give up aspiring for a more robust "evenhandedness" altogether. In the scientific community, as in other institutions (Chapter 2), there may be contestation about the values to be expressed in its practices with different articulations coming to the fore at different times as unavoidable tensions are played out in different ways. The modern scientific tradition has developed in mutually reinforcing interaction with the modern values of control; at the same time it has insisted that power (whether exercised with overt force or covertly through the normal functioning of dominant social institutions) should not define the scope of scientific inquiry, but that in principle none of the possibilities of nature and human nature lies beyond the scope of scientific investiga-tion. There is tension here, intensified when considering whether "science" should be defined by inquiry conducted under the materialist strategies or also include any systematic empirical inquiry (Chapter 5). Further tensions can be seen when science is linked with "progress": science serves the fuller manifestation of the modern values of control, but it also stands opposed to dogma and ideology wherever they are to be found, not just in pre-Enlightenment times and societies. In informing the projects of "progress," science promises soundly grounded knowledge not only about means to ends and about what is technologically possible, but also about the consequences and side-effects of technological implementations and alternatives to them – that is part of "evenhandedness." Thus, for example, in the case of the seed, "evenhandedness" is ignored when scientific applications further the interests of agribusiness in a context where "no lost possibilities" (since not grounded in empirical investigation) functions in an ideological role. I propose N as an expression of *neutrality* aiming to capture this alternative tendency carried by the tradition of modern science.

I endorse N as a value of the practices of science. Endorsing N goes hand in hand with endorsing I. Clearly N presupposes I; if theories are not accepted in accordance with I the question of *neutrality* does not arise. In addition, *ceteris paribus* the degree of expression of I in scientific practices is increased as they express N more fully. Research expressing N provides the range of theories needed for testing the manifestation of the cognitive values in theories against the highest standards, and it ensures that a greater variety of theories can be tested in such a way that they can become accepted (or rejected) in accord with I. The high expression of N is, thus, a condition for I to be expressed widely throughout the range of systematic empirical inquiries. While I endorse N, I am well aware of the enormous practical difficulties confronting its fuller expression, for that would require fundamentally reconceiving and reorganizing the institutions of science. I endorse it because human flourishing, in as many dimensions and forms, and for as many people as possible, remains my point of reference. Human flourishing is enhanced when actions can be informed by sound empirical knowledge (wherever possible by posits accepted in accordance with I), so that it would be well served if there were available a variegated body of knowledge which provided, for each viable value complex, significant posits that were able to inform activity within any one of them – with the horizon that all adhered to viable value complexes might become socially sustainable.

Neutrality *is not "neutral"*

So my endorsement of N rests ultimately on a value, moreover one in tension with the modern values of control – N itself is not "neutral"; it does not command universal adherence. It articulates a kind of *neutrality* of scientific practice which cuts across value complexes including those that contain the modern values of control, since the inquiries it refers to incorporate those conducted under the materialist strategies, and so they produce theories that are significant where the modern values of control are manifested. But where multiple strategies are in play, the conditions that sustain inquiry under them would be likely to curtail the amount of research conducted under the materialist strategies. Such research would be appropriate and sufficient where control is a subordinate value, but generally not where the modern values of control are adhered to. The interests of the modern values of control are not generally well (pre-eminently) served by scientific practices that aspire to the fuller manifestation of N.

The character of the *neutrality* of scientific practice cannot be settled by cognitive and methodological considerations, but ultimately by one's fundamental values. These values, in turn, are items in value complexes which also have presuppositions. A presupposition of the values connected with both the grassroots empowerment and feminist value complexes (Chapters 8 and 9) is that a variety of distinctive modes of human flourishing are possible and that they could be realized without causing the kind of human, social and ecological devastation that has come in the wake of modernizing "development." Value complexes that include the modern values of control have been articulated with various presuppositions, for example, that human flourishing is uniquely enhanced within them, and that there are no possibilities (other than massively destructive ones) for the foreseeable future outside of social forms in which they are manifested (Lacey 1997c).

Clearly none of these presuppositions (or their negations) at present can be accepted in accordance with I. They are not included in the current stock of soundly accepted theories and posits; they express anticipations, expectations, hopes, faith and the like. They also are about the possibilities of nature and human nature, so that, in principle, empirical investigation can bear on them in the long run, provided that it is conducted under multiple strategies.[16] Such investigation, whose practices would aim to express N, is appropriate not only if we desire to explore the range of ways in which flourishing lives can be lived, but also if we desire to hold all claims of the possibilities of nature and human nature – especially those pertinent to value complexes of contemporary social and political significance, and to the controversies among them – to consistency with the long-term outcomes of empirical investigation. Where scientific practices aim only to express (variants of) N'', such presuppositions will remain in principle outside of long-term empirical purview; and the argument for the presuppositions of the modern values of control will remain derivative to adhering to these values. Ironically, then, I appeal to *neutrality* in critique of the mainstream tendencies of modern science.

I have argued that N is worth endorsing, but also indicated the great difficulties confronting its fuller expression. In matters pertaining to values, moving from the outcomes of argument to action may be undermined by the unavailability of the material and social conditions necessary for the action to proceed. It is difficult to figure out what to do when one endorses N, though we find anticipatory pointers in Chapters 8 and 9. The situation is different with N'', for there are institutions well entrenched in contemporary society in which research conducted under the materialist strategies has a well established tradition, and in which the modern values of control are highly manifested and embodied. Thus the path to equilibrium through "adjustment" leads readily to engaging in research under the materialist strategies; whereas endorsing N arises in the paths of "creative marginality" and "transformation from below" (Chapter 2). In this way, the widespread adherence to N'' can be explained, even if – as I think – the argument favors adherence to N.

A further difficulty also arises: how to identify clearly what is to count as "science," "scientific" institutions and practitioners of "science" when the compass of what is to be included under "science" is expanded. Again, it is not the use of the word "science" that is in dispute; I want to include research conducted under the materialist strategies as one (very important) instance of systematic empirical inquiry, other instances of which include that conducted under the grassroots empowerment and feminist strategies (Chapters 8 and 9). The collected body of systematic empirical inquiries (carried out under a variety of strategies) and their theoretical products are not an integrated, unified whole, and many of these inquiries do not (in actual fact) even interact with one another. Regardless of the strategy under which a theory (posit) is generated and consolidated, what is important (for present purposes) is whether or not it is or (in principle) could be acceptable of some domain of phenomena in accordance with I.

Endorsing N leaves it open that particular investigators (and groups of them) work exclusively under particular, favored strategies, and do so because of the strategies' mutually reinforcing interactions with their particular, adhered to values, provided that they do not (cognitively) delegitimize research under all other strategies; and remain open, in principle, to rethinking their position in the light of the outcomes of the alternative research projects that they would tolerate being carried out within the scientific community. Endorsing N does not free us from controversy, including that of the most intractable kind, though it may permit people to enter into controversy entertaining more modest and less self-interested claims.

Summary

So, is science *neutral?*

Is science – either that practice conducted largely exclusively under the materialist strategies, or the collected body of systematic empirical inquiries – a highly rated object of value for all value complexes? – No.

Does accepting theories, in accordance with *I*, rationally commit one to particular value judgments? – I have not challenged this.[17]

Can theories, accepted in accordance with *I*, cognitively reinforce or undermine the values one holds? – Yes.

Are theories, soundly accepted in accordance with *I*, generally significant for all viable value complexes? – No.

Is the *neutrality* of scientific practice well manifested in actual fact? – N'' is; *N* is not; and no version of *neutrality* that includes a robust version of "evenhandedness" is.

Is *neutrality* a value to be adhered to in on-going scientific practice? – If the modern values of control are adopted, N'' (but not *N* insofar as it goes well beyond N'') tends to be a highly rated value; otherwise, it is subordinated to some other values. If the value of nurturing the many and various modes of human flourishing is adopted, *N* is a highly rated value and efforts to develop multi-strategy alternatives to the materialist strategies will be endorsed. Either way, taking *neutrality* to be a highly rated value springs not from cognitive and methodological considerations alone, but also from commitment to a particular social value.[18]

AUTONOMY

Autonomy was proposed (Chapter 4) as a view about features of scientific practices, and the processes of their conduct, desired for the sake of furthering the manifestation of *impartiality* and *neutrality*. I offered provisionally the following formulation (abbreviated version):

A' 1   Scientific practices aim to gain theories that are accepted in accordance with I' and whose acceptance accords with N'.

   2   They are conducted without "outside interference" by the scientific community which: a) defines its own problems, etc,; b) has unique authority with respect to matters of method, etc.; c) determines who is admitted into the scientific community, and what counts as competence and excellence; d) shapes scientific education and scientific institution; e) forms its members in the practice of the "scientific ethos"; and f) exercises its responsibility to the public fully by acting in accord with items a)–e).

   3   The scientific community conducts its investigations in self-governed institutions which are free from "outside interference", but provided with sufficient resources in order to conduct its investigations efficiently.

As with I′ and N′, it is presupposed in A′ that scientific inquiry is conducted virtually exclusively under the materialist strategies; so, minimally, it must now be revised in order to fit with *I* and *N*. Revising A′ turns out to be an elusive quest, leaving *autonomy* at best a distant and tension-filled ideal.

Autonomy is subordinate to *impartiality* and *neutrality*. Thus, a satisfactory thesis of *autonomy* will include (leaving open for now what the other items might be):

A 1 The practices of scientific (systematic empirical) inquiry aim to bring about in themselves fuller expressions of *I* and *N*.
....

A(1) presupposes that the fundamental objective of scientific (systematic empirical) inquiry is to gain understanding of phenomena (O) and this includes the encapsulation of their possibilities. O, however, cannot be pursued directly but only *via* following particular approaches (O$_i$), which are defined by the adoption of particular strategies (Chapter 5), and which enable the encapsulation of possibilities of particular kinds. Adopting the materialist strategies represents following one particular approach (O$_1$), one among (potentially) many, the one that enables the encapsulation of the material possibilities of things. It is adopted almost exclusively because of its mutually reinforcing interactions with the modern values of control (Chapter 6). Then, we see that A′(2) (and (3)) cannot be sustained, for the confining of research to that conducted under the materialist strategies is intelligible only in terms of the connection with an "outside interference," the interests of the modern values of control – an "outside interference" already, as it were, built into A′(1).

A′ is not responsive to the need to *choose an approach* to follow, and so it does not recognize the place where values can play their proper and essential role in scientific practices. "Outside interferences" – such as values – must play a role in the process of scientific inquiry at the (logical) moment when strategies are adopted. A′ can be rendered intelligible by interpreting "outside interference" as "outside interference other than that implicit in the mutually reinforcing interaction between the materialist strategies and the modern values of control." Under this interpretation one might hold that A′ is actually reasonably well expressed in the practices of modern science (despite notorious departures from it), especially if allowance is made for the "compromises" discussed in Chapter 4 – so much so that its widespread endorsement among the members of the scientific community and scientific institutions may be an important causal factor in the inadequate expressions of *I* and (especially of) N in contemporary scientific practices. But, no matter how much it may be actually expressed in scientific practices, A′ makes no general claim to being adhered to, for it applies only to practices that generally can further the manifestation of N″, but not N. Endorsing A′ (with obvious minor modifications) makes sense, as the expression of a value of scientific practice, where N″ is endorsed, thus where the modern values of control are adhered to – but not necessarily among the general public, and certainly not among those (Chapter 8) who are attempting to develop systematic empirical (scientific) approaches that are not linked with the modern values of control.

Not to endorse A′ is not the same thing as to reject it outright. My objection is that it rests on misidentifying O with O_1. But O_1 does express the objective of an important approach to science; so while there is not a general interest in its pursuit or in the unqualified endorsing of A′, it is important that the results obtained under it accord with I. When O_1 is subordinated to the aim expressed in $A(1)$, how can we ensure that its theoretical products accord with I? How should scientific practices be conducted so that values play their role at the admissible moment, without opening the door to any "outside factors" (dogma, ideology, government, corporations, popular opinion) playing any role that might be chosen for them by those with the power to do so? I propose that O_1 should be pursued according to a thesis of *localized autonomy*: within certain bounds – defined, I suggest, not exclusively by self-governing scientific institutions, but in the course of inclusive democratic deliberation so as to be responsive to N – the items of A′(2) should be endorsed, with virtues listed in the "scientific ethos" multiplied to include "openness to the results, accepted in accordance with I, of research conducted under other strategies," "sensitivity to the distinction of levels – of strategy and theory – and to the legitimate role of values being confined to the former level," and "interest in discerning and submitting to empirical investigation, where possible, any posits that are key to important current value disputes."[19] I will not explore this notion of "localized autonomy" further.

In the context of O, and the recognition of multiple approaches to furthering it (each linked with particular values), it is difficult to identify items that might be added to $A(1)$ in order to state a generally defensible thesis of *autonomy*. The interest in the fuller manifestation of I and N, however, remains. How must scientific research be conducted, and in what kinds of communities, in order to gain more theories (posits) accepted in accordance with I in the course of inquiries that increasingly express N? In view of the preceding discussion of *neutrality*, I suggest, pending their further testing in relevant empirical investigation, that at least the following conditions need to be satisfied (cf. Longino 1990: 76). First, research is conducted under a range of strategies, which are sufficiently varied to permit that, for each viable value complex currently adhered to, some research is conducted under a strategy in mutually reinforcing interaction with it. Then research is being conducted so that (in principle) if it is successful, for each viable value complex, there will be produced significant theories, and in the course of its conduct, the presuppositions of all these value complexes, if at all possible, will be put to some measure of empirical investigation. In order to further the manifestation of N, values are not kept out of roles in research practices (or their role disguised); rather appropriate roles are provided for a variety of value complexes.

Second, the distinction of levels (of strategies and theories) is strictly adhered to, with the role of values limited to that of providing support for the adoption of particular strategies – delineating the general features of the possibilities of interest to be investigated and framing the context in which theories consolidated under the strategies are likely to be significant – and not permitted to intermix with the cognitive values in making judgments of theory acceptance and rejection. So that the distinction of levels (strategy and theory) can be respected, the practices of each community, engaged in systematic empirical inquiry, are granted a version of localized autonomy, the details of which still need to be worked out, parallel to that introduced for communities pursuing O_1.

Third, the research is conducted by a variety of interacting communities committed to systematic empirical inquiry. While a particular community may (and typically will) conduct research principally of a particular approach, there is sufficient interaction (and the institutional structures to ensure its being sustained) – framed by the objective O – that the theories accepted in accordance with I of each (when relevant) can be brought critically into the deliberations of the others. Then the standards of testing of all will be raised, and posits of importance to some value complexes will not be dropped from consideration prior to the appropriate investigation of them being carried out. Fourth, the items of the "scientific ethos" are expanded to include those listed on p. 250 in connection with the discussion of the localized autonomy of research under the materialist strategies.

Fifth, the matter of who (and the qualifications of whom) is regarded as a member of a "research" community admits of considerable flexibility, and is open to ongoing contestation. It may lack ready definition (in terms of, for example, qualifications), especially (Chapter 8) in view of there being approaches attempting to develop a dialectic between traditional knowledge practices and research conducted under the materialist strategies. Then, a "research" community may not be separable from a community engaged in a particular set of practices (for example, agricultural) – systematic empirical inquiry may be a part of a more general, on-going practice.

Finally, in part elaborating some of the earlier conditions, the community of communities engaged in systematic empirical inquiry needs to manifest "diffusion of power" (Longino: see Chapter 9). Diffusion of power, in principle, functions as a counter to the impact of values on scientific production present because science is produced mainly in institutions, such as universities and research institutes, which are an integral part of the dominant socio-economic system. These institutions often reinforce certain values, perhaps misidentifying them with items of the "scientific ethos," such as the primacy of the intellect, individualism, competivism, and virtually exclusive commitment to one's scientific discipline. They require of the scientist an intensive training, and then certain priorities of time and commitment. In these institutions, being a scientist is a "full-time" commitment, and family, social and other involvements must adapt to it; otherwise one will fall behind in the scientific competition. It involves a way of life that makes it difficult to coexist with lives that express certain other values, for example, those expressed through participation in programs for social transformation (except those of modernizing: Chapter 8). Moreover, it is a way of life that tends to be granted a sort of "epistemic privilege" (cf. Longino 1993), a privileged place for making authoritative judgments about what is and what is not possible. So widely is it accepted that modern scientific theories, developed under the materialist strategies, provide the best encapsulations of the possible; a privileged place which generates asymmetric relations, which are conducive to relations of domination between those whose practices are informed by the scientists' knowledge, and those who are expected to adapt their lives to conform to the imperatives of scientific knowledge. If one does not adopt such values, one will not be able to participate in the experimental and theoretical practices of modern science, and so one will not gain the experiences necessary to be an "expert" in the domain of scientific knowledge. But, if one does adopt them, one will most likely not be able to recognize the practices from which alternative forms of social transformation might develop. Adopting the values, then, will enable one to gain access to more and more material possibilities that might be realized through technological practices, but it may leave one in virtual ignorance of novel social possibilities (and the material possibilities that they might introduce) that actually existing reality may permit.

In the light of the dialectical connections that I have referred to, I wish to suggest – as an item for further discussion – that there is no way of life and there are no institutions, which enable one to explore *both* the fullest range of modern technological possibilities *and* the possibilities of social transformation that would adequately serve the poor majorities of the impoverished countries and the interest of expanding the capabilities of human agency in general.

Adopting the social value of diffusion of power, therefore, has far-reaching implications. I have defended it in terms of its potential to serve cognitive ends: to produce higher standards for estimating the degree of manifestation of the cognitive values, and to gain access to possibilities that cannot be encapsulated under the materialist strategies (Chapters 8 and 9). As a social value it can also be defended in terms of its contribution to social justice. I think it important not to dissociate the two defenses. When they become dissociated and the one from social justice prioritized, furthering diffusion of power can also lead to a lowering of cognitive standards and the failure to gain the kind of knowledge that is needed to inform the projects of social transformation (Chapters 4 and 9).

I am unable to bring these conditions together into a sharp thesis that might replace A$'$, that could be endorsed on similar terms to N. There are several reasons for this. In the first place, the set of approaches that might be adopted under the general umbrella of O is too ill-defined (though certain of its members can be readily identified: Chapters 8 and 9), as also is the set of research communities to be considered included in the body of interacting research communities. Second, the value of research meeting conditions like these is contested by those who adhere to the modern values of control who aspire only to manifest N'' and not N so that, for lack of social and material conditions and the presence sometimes of outright opposition backed by power, numerous potential approaches to science are not entertained or sometimes even conceived. This renders it even more difficult to hope to identify concretely a body of communities whose research, collectively and interactively, might be sufficient to ensure the fuller expression of N. Third, endorsing N and seeking ways to enhance its expression, are likely to lead one to challenge the practices of any actual community (of communities) engaged in scientific research for lacking significance for certain specified value complexes; and one may need to back this challenge with concerted political action to bring the community's practices more into accord with N.

In such a situation, *autonomy* – other than pragmatically and democratically worked out "localized autonomy" – has little meaningful, short-term relevance. We can only conjecture about what a community of communities of research would be like for which we might desire to endorse a form of *autonomy* – that is, a community sufficiently specified and institutionalized, whose on-going conduct sufficiently reflected the six conditions listed on pp. 251–2. For the long run, however, such conjecturing is not idle for, given commitment to the multiplicity and variability of forms of human flourishing, N is worthy of much fuller expression. If eventually there are created institutions that serve effectively to further it, then a version of *autonomy* that is consistent with the six conditions may be able to be stated and defended. But that remains conjecture. I add the further conjecture that it would further the traditional goals of the university if it committed itself to the furtherance of N.

11 Conclusion

Is science value free? Do the practices of scientific inquiry express the values of impartiality, neutrality, and autonomy? *Impartiality*: that a theory is accepted (of a domain) if, and only if, in relation to the appropriate empirical data, it manifests the cognitive values to a high degree according to the most rigorous available standards. *Neutrality*: that, for any viable value complex, there are theories (accepted in accordance with *impartiality*) which may be applied so as to further significantly the manifestation of the values that constitute it. *Autonomy*: that scientific practices proceed, and scientific communities and the institutions that support them are constituted, for the sake of furthering manifestations of *impartiality* and *neutrality* without direction ("interference") from factors other than the data, the cognitive values and the ingenuity of the practitioners, in particular without direction from (non-cognitive) values.

Impartiality and *neutrality* have presuppositions: *impartiality* that cognitive values are distinct and distinguishable from other kinds of values; *neutrality* that accepting a theory is logically consistent with making any value judgments, and that accepted theories leave open a range of viable value complexes. I have not questioned these presuppositions. I have defended endorsing the three values against common criticisms that they are not always actually well manifested: not all theories actually accepted in the scientific community accord with *impartiality*; most actually accepted theories (contrary to the aspiration of *neutrality*) serve to further the manifestation of some value complexes (those containing the modern values of control) much more than that of others; and the claim of *autonomy* must contend with the numerous compromises into which the scientific community enters with the values of those institutions which provide for its material and social conditions. For even when the three values are not manifested highly in actual fact, they may still be expressed in the practices of scientific inquiry. Then, the key question becomes: Are the practices of scientific inquiry so constituted that (when properly conducted) there is progressive movement in the direction of the higher manifestation of *impartiality*, *neutrality*, and *autonomy*? If so (for all three values), then "science is value free" would be well grounded.

The fundamental objective of scientific inquiry is to gain understanding of phenomena: to chart and explain their workings and features, to encapsulate the possibilities (including novel ones) that they allow, and to discover how some of the hitherto unrealized ones may become realized. The cognitive values designate those characteristics of theories (and their relationships with empirical data and other theories) which indicate that they serve to fulfil this fundamental objective with respect to some domains of phenomena. The objective of gaining understanding, however, provides by itself no direction to scientific inquiry. In order to pursue it, it is necessary to follow a particular *approach* to inquiry, where an approach has the objective to investigate possibilities of a specified kind. Following a particular approach involves adopting *strategies* which constrain the kinds of theories entertained and select the kinds of empirical data which those theories should fit. A strategy gives direction to research; and the adoption of a strategy gains support, from which the grounds to adopt it cannot be separated, from its mutually reinforcing interactions with (moral and social) values. Sometimes a strategy will be adopted deliberately and consciously because of its relations with values; though, more typically, certain strategies frame practices of inquiry that are solidly entrenched (so that adopting them seems not to require rational justification) because the values with which they bear reinforcing relations are deeply manifested and embodied in prevailing social institutions – as the materialist strategies are entrenched in view of their mutually reinforcing relations with the modern values of control.

Since engaging in scientific inquiry (which I construe as any form of systematic empirical inquiry) requires adopting a strategy, and adopting a particular strategy becomes intelligible only in view of its mutually reinforcing relationships with values, *autonomy* cannot be sustained. Values pervade, and must pervade, scientific practices and (in significant part) account for the direction of inquiry and for the kinds of possibilities attempted to be encapsulated in theories.

The link with values becomes further apparent when it is observed that theories that become accepted in the course of inquiry conducted under given strategies generally are applicable in ways that serve to further especially the values that have reinforcing links with the strategies. Theories accepted under the materialist strategies, for example, on application serve especially well interests framed by the modern values of control. Since the materialist strategies are adopted virtually exclusively in modern scientific practices, this indicates that these practices do not actually serve the fuller manifestation of *neutrality* well. Does this mean that *neutrality* cannot be sustained? I have suggested that it does not, that *neutrality* can be sustained as a value of scientific practices provided that there are in play within scientific communities (the body of communities in which systematic empirical inquiry is conducted) a multiplicity of strategies, among which the materialist strategies may have an esteemed place – as in the formulation, N (Chapter 10). There is good reason, I have also suggested, to endorse N; but then the ground of *neutrality* is not in a methodology based in strategies which deploy a lexicon that is devoid of value terms, but in the value of the social sustainability of a variety of forms of human flourishing (not limited to those that are implicated in the modern values of control). On the other hand, N can become manifested more fully only if there is significant change in the conduct of scientific practices (away from the virtually exclusive adoption of the materialist strategies) and the character of their supporting institutions, and – in light of the mutually supporting relationships of strategies with social values – this can only happen if there is change in the values that are predominantly manifested and embodied in social institutions.

N would be endorsed, I suspect, only among those who adopt strategies incompatible with the exclusive role of the materialist strategies, and who question the modern values of control on the ground that their manifestations and embodiments threaten the social sustainability of valued forms of human flourishing. It is not endorsed in the mainstream of modern scientific inquiry. Moreover, practices in which the materialist strategies are adopted almost exclusively do not and cannot come to express N more fully. They are linked too closely with "progress," with furthering the modern values of control. But they do retain a residue of *neutrality*; accepted theories may be applied to further significantly the interests of all value complexes that contain the modern values of control. For those who think that the trajectory of the future has been set and that, in fact, there are no significant realizable possibilities for the future outside of value complexes that contain the modern values of control, this residue is enough of *neutrality*; for them N would represent either idle rhetoric or illusory aspiration. Whether one endorses N or the residue of *neutrality*, then, depends on the values one adheres to and on one's anticipations of future possibilities; the matter depends not on issues of scientific methodology, narrowly conceived, but on substantive value controversy and disagreements open to investigation in the social sciences.

What of *impartiality*? That values are implicated in the adoption of the strategies under which theories are generated, developed and consolidated permits (in principle) that theories can be accepted (of various domains of phenomena) solely in the light of the empirical data, other accepted theories and the play of the cognitive values. Numerous theories, developed under the materialist strategies, have been accepted in accordance with *impartiality* – though actual departures from it are not uncommon, especially when certain kinds of theories are rejected, not because of their cognitive failures, but because they do not fit the materialist strategies. The practices of scientific (systematic empirical) inquiry are constituted so as to permit fuller manifestations of *impartiality* in the making of theory choices. Thus, *impartiality* remains a defensible and, I believe, obligatory value of scientific practices. It is a value rooted in the very objective of gaining understanding, a requirement of being able to separate the genuinely possible from the merely conceived or desired to be possible.

In various ways understanding informs our activities and social practices; it informs means to ends, the possibility of desired ends, and the presuppositions that lie behind the legitimation of our projects. The value of the social sustainability of a wide variety of forms of human flourishing supports that the fuller manifestation of *impartiality* is a value. *Impartiality* is manifested more fully when (1) increasingly the theories that are actually accepted in the course of scientific inquiry accord with it, and (2) the range of phenomena of which theories are accepted in accordance with it continues to expand – moving towards the ideal in which our activities (both their concrete conduct and their legitimation) are informed wherever possible by understanding that accords with it. Conducting inquiry almost exclusively under the materialist strategies has produced a remarkable number of theories accepted of numerous and various domains of phenomena (the range of which continues to expand). It does not permit, however, moving towards the ideal, for it lacks the means to investigate whole classes of possibilities, those available to phenomena when we do not abstract them from their human and social contexts, for example, to investigate the question of whether there are significant realizable possibilities for the future outside of those that serve the interests of value complexes containing the modern values of control. Manifestations of *impartiality* can only move towards this ideal when inquiry is conducted under a variety of strategies. Only when diverse values lead to the adoption of a variety of competing strategies – rather than when it is attempted "to keep values out of science," but when in fact the attempts become a disguise for science being conducted under one kind of strategy and thus under the influence of one set (no matter how "universally endorsed") of particular values – are the conditions in place for the kind of manifestation of *impartiality* that could approach the ideal. When, and only when, these conditions are in place, can we also expect that fuller manifestations of *impartiality* will be accompanied by fuller manifestations of *neutrality* (N).

Is science value free? It is not, since *autonomy* cannot be sustained. But it is not illuminating to sum this up with the slogan: "Science is not value free." I have been more concerned to articulate the positive ways in which values may (legitimately) play a role in science, and with how these are compatible at least with *impartiality*, a core component of "science is value free"; and to make clear that their legitimate role actually requires commitment to *impartiality*, for the adoption of a strategy – though its rationale comes from relations with values – is not justified in the long run unless it leads to the development of theories that become accepted in accordance with *impartiality*.

Values, I have said, pervade and must pervade the practices of scientific (systematic empirical) inquiry, the practices whose objective is gaining understanding of phenomena. That is because scientific inquiry is conducted under strategies, the adoption of which is explicable in terms of their mutually reinforcing interaction with particular values. Values have a proper role to play at the moment of the adoption of strategies. But their legitimate role does not extend to playing alongside the cognitive values in making theory choices.

The separation of moments – adopting strategies, accepting theories – is essential, with values playing their role only at the moment of adopting strategies. It permits the retention of *impartiality*; and makes it possible to anticipate that conducting research under a variety of strategies might lead to fuller manifestations of both *impartiality* and *neutrality* (N). It also helps to explain why inquiry conducted almost exclusively under the materialist strategies lacks *neutrality* (N) and manifests at most its residue. It also allows us to identify mechanisms (when values, consciously or not, come to function alongside the cognitive values) that explain how *impartiality* can actually be departed from in scientific practice. The separation of moments underlies a synthesis of the genuine achievements of scientific inquiry conducted under the materialist strategies, and what is sound in the many contemporary forms of criticisms of science; without having to limit scientific practices to those of the former (with its often implicit commitment to materialist metaphysics) or to embrace the radical relativism often associated with the latter

The account enables us to pick out what is and what is not defensible in "science is value free"; (partly) to explain why *neutrality* (N), although defensible, is not widely endorsed, why *impartiality* is not always well manifested in actual fact, and why *autonomy* seems defensible to those who do not accept that there are questions about choice of strategy to be addressed; and to characterize in broad terms how inquiry must be conducted if there is to be fuller manifestation of *neutrality* and of (an ideal of) *impartiality*. It also provides a setting in which value-based criticisms of current scientific practice can be made. The bulk of modern scientific inquiry is conducted under the materialist strategies, and these interact in mutually reinforcing ways with the modern values of control. How would the practices of scientific inquiry have to be re-constituted so that its strategies come to interact in similar reinforcing ways with the value of the social sustainability of the widest possible variety of forms of human flourishing? The discussion of the "grassroots empowerment" and "feminist" strategies provides some initial hints. With the separation of moments in my account, this becomes an intelligible question. It is also one of the urgent moral questions of our day.

Notes

1 Introduction: the idea that science is value free

1 As commonly held among logical empiricists (Feigl 1961: 15) and critical rationalists (Popper 1972: 29).

2 Descartes and Newton played the key roles in consolidating the fusion of the two ideas in the tradition of modern science.

3 Implementing this proposal involves presuppositions about authoritative testimony (Coady 1992) and the availability of appropriate channels of communication. Deepening it requires consideration of "Rudner's argument" (Chapter 4).

4 See Cupani (1998) for a defense, and Hull (1998) for a challenge.

5 Arguments of this kind are often part of "feminist empiricism" (Anderson 1995a; Nelson 1996).

6 My usage differs from that of, for example, van Fraassen (1980).

7 Almost all of them have been elaborated by Weber (1949). He also maintains that commitment to "science is value free" does not imply moral indifference, but is a presupposition of sound moral thinking and action.

8 Roughly, the fusion of the Galilean and Baconian ideas under discussions rests upon yet another Baconian idea, the valuing of "utility."

9 "Strategy" serves many functions similar to Kuhn's (1970) "paradigm," "disciplinary matrix" and "structured lexicon" – and, to some extent, Lakatos' (1970) "research programme." I do not use Kuhn's notion, because of its changes over the years, to avoid entering into questions of Kuhnian interpretation, to keep a distance from some of the connotations associated with his terminology, and so that I assume full responsibility for the implications of my use of the notion. Nevertheless, my indebtedness to Kuhn is great; "strategy" is certainly an intellectual descendent of "paradigm." Note also that there are other dimensions of strategies (and paradigms) that I will not develop in this book, e.g., concerning the provision of guidance for the deployment of the auxiliary hypotheses that mediate between theory and data.

10 Holist epistemologies, that do not readily permit differentiation of levels, often underlie criticisms of "science is value free."

2 Values

1 This chapter is a shortened version, with some changes, of parts of an article I co-authored with Barry Schwartz (Lacey and Schwartz 1996). In it we analyzed how values can be considered simultaneously to be objects of cognitive and moral evaluation, and objects of social and psychological explanation. Further details on many points raised in the text can be found in the article.

2 On the plurality of values and the variety of appropriate relations with objects of value, see Anderson (1993).

3 In this section, I draw freely upon Nerlich (1989) and Taylor (1985).

4 Values discourse becomes deformed when "values" become separated from their other modes, especially when holding values becomes identified solely with their articulation (Lacey and Schwartz 1996).

5 We also say that "X values *v*," where *v* is "an object of value" for X. Usually "X values *v*" fills in for an expression like "X values that ø be characterized by a certain type of relationship with *v*." Each object of value is associated with specific relationships (Anderson 1993).

6 Graham Nerlich (correspondence) has strongly questioned my tying of evaluation to future orientation.

7 Where one discerns, in interpretative inquiry, that a value is manifested fairly constantly in a person's behavior, but the person displays no reflectiveness about its desire and belief dimensions, and displays no particular commitment to narrow the gap, I will refer to the person as merely *having*, rather than *holding* the value.

8 "Holds" as distinct from simply "has" a value (see previous note).

9 McMullin (1983) and Nagel (1961) refer to all three kinds of these judgments as "value judgments." McMullin calls the first two kinds of judgments "valuing," and the third "evaluating." Nagel (1961) refers to them respectively as judgments of "appraising" and "characterizing."

3 Cognitive values

1 Some of the beliefs that inform one's actions may remain unevaluated and not become objects of deliberation (or even recognition) when they derive causally from: mechanisms for the inculcation of ideology, the absence of social institutions in which alternative views are articulated and critically discussed, prejudice, habit, brainwashing, indoctrination, desire to conform, expediency, opportunism, unreliable authorities, threat, ignorance of alternatives, exigencies and limitations of experience, unsound argumentative moves, cognitive errors, behaviorist reinforcement, projections of class interests or personal desires, and simple unreflectiveness or other personal vices.

2 We do not always have means available to resolve the contradiction. There is a manner of speaking in which people say "p is true for me, but apparently not for you." I take this to mean something like: "p is sufficiently well supported to warrant informing activities connected with my attempts to manifest more fully my most fundamental values; but apparently not so with yours." It may be that most of the beliefs that inform our activity are of this type – then we may be mired irremediably, in "*de facto* relativism." Even if this is our predicament, it does not provide a reason not to be responsive to argument and evidence.

3 At times the interpretative activities of identifying the cognitive values manifest in beliefs and the values manifest in behavior will be intertwined. When considering certain types of beliefs, their having a specified causal origin – for example, origin in appropriate experiences – can be a highly rated cognitive value. This point is heightened when we take note that most of our beliefs rest upon the testimony of others (Coady 1992), so that there will be cognitive values of the type "testified to by a reliable authority." Furthermore, reliable observational reports, especially of certain social phenomena, may require observers who manifest particular values, since the phenomena may be opaque to those who lack a certain kind of moral sensibility (Lacey 1997c).

4 Or, at least, for each kind of object of investigation, there should be a correct set of cognitive values to frame the investigation. The possibility that, to some extent, cognitive values may be "relative to" the object of investigation will not be explored in this book.

5 The controversy over creationism partly concerns the list of cognitive values to be embodied in public educational institutions.

6 McMullin (1983) points out that, ironically, the attempts of the positivist-influenced traditions to reduce cognitive value to the outcomes of rule-governed operations all covertly traded on subtle cognitive value judgments. The application of inductive rules presupposed "judgments" – which are not rule-governed – about "curve-fitting, extrapolation and estimates of relevance"; and in order to characterize "basic statements," Popper (1959) appealed to *conventions* whose deployment involved not rules, but such (cognitive) value judgments as "easy to test" and "likely to gain agreement." Finally, Carnap (1956) based the choice of theoretical language (sensationalist, physicalist, etc.) on such values (cognitive, pragmatic?) as "efficiency, fruitfulness and simplicity."

7 The way of looking at things can be traced back to Kuhn (1977) with anticipations in the "Postscript" to Kuhn (1970); it has been most richly developed by McMullin (1983, 1993, 1996): see Lacey (1997b).

8 Sometimes, cognitive values are called "epistemic values" or "epistemic virtues" (McMullin 1983, 1996), or "epistemic utilities" (Hempel 1981). I am contrasting cognitive values with relations (between theories and data) which, like proof in mathematics, obtain only if the relata are connected in a way that can be established by a set of formal rules. In another sense of "rule" (for example, Eldridge 1997), rules are followed whenever one accepts a theory, for example, the rule: accept only theories that manifest the cognitive values highly; and scientific practices are structured by various "rules": orderly, somewhat codified, and normative constraints.

9 The list simply draws together items that have been proposed in the literature: especially Kuhn (1977) and McMullin (1983; 1993; 1996); also Ellis (1990), Laudan (1984), Longino (1990), MacIntyre (1977), Newton-Smith (1981) and Putnam (1981; 1987; 1990).

10 Van Fraassen (1980) defines empirical adequacy, in the context of the semantic model analysis of theories, as: T has sub-models that are isomorphic with sets of observable phenomena of relevant domains. These sets cannot be identified with sets of items of E, since E consists of actual data. However, it seems to me, we have reason to *believe* that T is empirically adequate in Van Fraassen's sense if, and only if, (1)–(4) are satisfied.

11 For queries about the last clause see Brusch (1989).

12 Or *non-triviality*: not all statements well formed with the categories of the theory are entailed within the theory (da Costa and Bueno 1998).

13 Salmon (1966) regards "analogy" and some other items cited here as factors that are relevant to the assessment of "prior probabilities" for purposes of Bayesian calculations.

14 See Lakatos (1970) on various senses of "*ad hoc*."

15 I am not aware of the issue of standards having been directly addressed in the literature on cognitive values. I offer here a preliminary discussion that needs elaboration and critical interaction.

16 In Lacey (1979) I also argue that the data considered by radical behaviorism do not meet the representativeness standard.

17 For example, empirical investigation leads researchers in medical science to regard double-blind testing as more reliable than other methods (Anderson 1995a; Laudan 1984).

18 This standard functions in close concert with *relevant to the critical confrontation of competing theories* (above), reflecting Feyerabend's (1965, 1975) argument that theoretical conflict can be an instrument for increasing empirical adequacy. In practice, it can be very difficult to interpret because of the well known thesis of the underdetermination of theories by data.

Note also that a certain explanatory ideal (e.g., mechanism, holism, organicism, intentionality) may function as a standard, though often it may be difficult to separate its role from that of strategies (introduced in Chapter 4; see also Chapter 5).

4 Science as value free: provisional theses

1 The notions of "accepting," "applying" and "significance of" a theory were introduced in Chapter 1, as was the terminological convention to distinguish "cognitive values" from (other kinds of) "values."

2 Elsewhere (Lacey 1997b, 1999a, 1999b) I have called them "*materialist constraint / selection strategies*" in order to use a terminology which always puts the two roles of strategies at the center of attention.

3 The Galilean idea (Chapter 1) is metaphysical; adopting the materialist strategies follows from accepting this idea (materialist metaphysics). My terminology, "materialist strategies" derives from this (see McMullin 1999 for an objection). Without denying the historical importance of the metaphysical idea, adopting the materialist strategies can be de-linked from it. In this chapter, I treat the materialist strategies as a commonplace of scientific practice. Then (Chapters 5–7) I argue that the ground of their almost exclusive use lies in links with values, rather than metaphysics.

4 For further clarification see Lacey (1999a). Some fields of science do not display the same kind of commitment to use of the materialist strategies. We may discern a spectrum of cases, ranging from the thoroughly materialistic to the "historical" (evolution, geology), to the naturalistic (natural history, taxonomy), to ecology, to the human and social sciences – so that in what is generally identified as "modern science," a range of strategies are in play.

5 Some comments: First, if one were to accept a rule-governed account of the criteria of cognitive value, item (1) of I' could be dropped, and (2) would be replaced by something like: "T is accepted of D if, and only if, the degree of its cognitive value, as calculated through application of the rules, is sufficiently high. ... "   Second, agreement in the scientific community does not suffice

for the acceptance of T, according to I'. But, given that the standards include having adequately responded to criticisms, a significant amount of agreement is a necessary condition for sound acceptance. Third, values other than cognitive values, may be manifested too in T, alongside the cognitive values, but this is irrelevant to soundly accepting T. Insisting that T manifest the cognitive values according to the highest available standards is a means for ensuring that such values do not play a (covert) role in T becoming accepted.

6 (R) is clearly a condition upon the legitimacy of any practical application of T. Many commentators have held that it is only that.

7 Questioning, of course, often does come to an end, especially in the natural sciences where it is uncontroversial that (R) holds widely.

8 See the discussion of "no lost possibilities" (Chapter 8; Lacey 1997c).

9 The notion of scientific knowledge as a public good is under strong challenge at the present time, especially in the light of the intertwining of biotechnology research with business interests that has been reinforced by extensions of intellectual property rights to cover, e.g., genes. *Neutrality* may well be rapidly disappearing as part of the self-understanding of the scientific community. So, too, may the link between the materialist strategies and the modern values of control (Chapter 6) – leading to scientific research strategies progressively becoming linked piecemeal with a variety of competing economic interests.

10 There is another view (which I will not discuss in this book) – call it "no value implicatures" – that might be included here: Accepting a theory in accordance with I' has no implicatures ("pragmatic implications" – Root 1993, drawing upon Grice) in the realm of values.

11 Since there may be theories accepted of domains with only the most remote connection with daily life and experience and practical activities, to the statement of "evenhandedness" should be added the qualification; " ... theories, accepted in accordance with I' *if they are applicable at all ...* ." I will assume this qualification to be in place here and in all subsequent revisions of "evenhandedness".

12 This argument is elaborated and clarified in Lacey (1997c).

13 "Consistent with all value judgments" is presupposed in the formulation of the other components of *neutrality*. It should be understood as containing a qualification. Earlier (Chapter 3), I suggested that it accords with an informal ideal of rationality that if one holds the belief that p then one negatively values any of one's actions that might be informed by ¬p; they are irrational. I interpret N'(1) (and all its successors) to be qualified by "with the exception: X's acceptance of T (of D) commits X rationally to valuing negatively any application X might make of negations of T (of D)." This qualification does not threaten "evenhandedness," for minimal rational commitments of this type are endorsed in all value complexes.

14 Another set of questions arises here too, for the issues of *autonomy* and responsibility of scientists are intimately connected with how scientists exercise their authority. Under what conditions "should" scientific claims be held by all, including non-scientists, on pain of the charge of "irrationality?" How should science interact with popular beliefs, outlooks and worldviews? How does science illuminate the realm of daily life and experience? Why, and within what bounds, should non-scientists accept the authority of scientists? Presumably, scientists, *qua* scientists, have at most cognitive and not moral authority. What happens when there is tension between them? In what ways does the authority of scientists, in actual practice, contribute to the entrenchment of

ideology? It is not uncommon for scientists, claiming to be speaking authoritatively, to affirm that science either presupposes or supports a metaphysical position, e.g., materialist metaphysics. In doing so, the claim goes beyond the evidence. Some forms of "anti-science" critique hold: that is the way science is; it always makes claims implicated in values!

15 Scientific institutions can undermine their own claim to autonomy in a number of other ways: for example, if they cease to be sources of authoritative testimony regarding what has been accepted in accordance with *impartiality* and *neutrality* – by tolerating or not investigating charges of fraudulent results, by only revealing results that serve special interests, or by using their authority so as to portray ideology as scientific knowledge. Where such things happen it may become appropriate to extend the areas of compromise to include also items (b) and (d) of A'(2). Finally, where the institutions in which scientific research is actually conducted turn out not to be genuinely self-governed (think of the research units of the tobacco industry), they can make no claim to be granted *autonomy* – freedom from public interference – on grounds connected with *impartiality* and *neutrality*.

5 Scientific understanding

1 Formal consistency with theories soundly accepted under the materialist strategies (or any other strategies) is a cognitive value.

2 See, for example, Laudan's (1984) analysis of the gradual rejection of the (supposed) cognitive value articulated in Newton's "*Hypotheses non fingo.*"

3 Shiva (1988) and Anderson (1995a) make the same point without the paradoxical rhetoric.

4 Theories accepted under the materialist strategies are unrivalled in having demonstrated high manifestations of the cognitive values. It does not follow rationally that we should always use soundly accepted theories of this kind to inform our actions (Chapter 3). If one aims for such objectives as the strengthening of one's cultural heritage, the empowerment of oppressed peoples, the overcoming of poverty, the enriching of relations among people, the strengthening of democracy, and the creation of ways of life that do not devastate the environment, then modern science (conducted under the materialist strategies) has (can have) little to contribute, except in a subordinate way regarding some of the technical projects that will be involved in acting for those objectives (Chapter 8).

5 Laudan (1984) has proposed that the identification and adoption of cognitive values can be rationally reconstructed in terms of a "reticulated model," which (paraphrased into my terminology) involves two-way interactions between each pair of the triad: {theories, scientific practices, cognitive values}. For example, cognitive values "justify" scientific practices, and scientific practices "exhibit the realizability" of the cognitive values; and theories (chosen in the course of scientific practices) and cognitive values "must harmonize." My account, in requiring that cognitive values be criteria of theory choice actually used in scientific practices, and that they be possibly manifested, incorporates Laudan's considerations. His account, however, does not distinguish between strategies and cognitive values, and does not involve significantly my fourth kind of consideration. He seems to hold that the objective of science is simply to gain theories that manifest highly the cognitive values that are currently adopted in scientific practice.

6 The argument deployed here has the same structure as that involved in Reichenbach's "pragmatic vindication of induction" (Salmon 1966); it is a priori. On the other hand, for example, the claim: "if we are to obtain empirically adequate theories in medical research meeting the standard of reliability, we must use double-blind testing procedures," is an empirical one. In general, procedures that are followed in order to bring about the high manifestation of cognitive values in theories need empirical vindication (Laudan 1984).

7 But see Chapter 4, note 4.

8 Made by John Clendinnen (in discussion). See also Chapter 7, note 18.

9 See Lewontin (1993) on the irreducible interaction between the biological and environmental causes of diseases.

10 And of the conditions provided for the research (Rouse 1987).

11 O′ is adapted from Ellis' statement of the realist's aim of science (borrowed from van Fraassen 1980), and O′ from his own version of the aim (Ellis 1985). (Recall – Preface – I will not discuss empiricist versions of the objective of science.)

Hempel (1983a) proposes that the objective of science may be put as: "seeking to formulate an increasingly comprehensive, systematically organized worldview that is explanatory and predictive." Acknowledging that this is very vague, he goes on to suggest that adopting a suitable set of cognitive values may be regarded as "attempts to articulate this concept somewhat more fully and explicitly." Then, the objective of science becomes to gain products that more fully manifest the cognitive values – so that they define the objective of science rather than gain justification from it (cf. note 5).

12 My account of multiple approaches has much in common with Tiles' (1985, 1987) view that scientific research unfolds in the context of agreements about empirical classificatory schemes and metaphysically derived constraints (which carry with them an explanatory ideal) on theories, and with Dupré's (1993) account of the disunity of science.

13 Another argument for adopting the materialist strategies exclusively goes: Physical science developed when it rejected strategies which deploy teleological categories, then biological science did the same, so we would expect that the same would happen in the human and social sciences. Another is sometimes used among those who adopt Bayesian approaches: The success of the materialist strategies confers high "prior probabilities" upon theories that are constrained in accord with these strategies. All such arguments involve what I call "the overgeneralization of meta-inductions" (Chapter 10).

14 Where I use "underlying structure, process and law," McMullin (1996, 1999) uses "causal structure of the natural world" (Lacey 1997b, 1999b).

15 It also follows that the problem with (for example) Lysenkoism was not in conception, but in implementation; its strategies were retained and practically adopted even though they failed (perhaps after an initial short-run appearance to the contrary) to generate theories that manifested the cognitive values highly.

6 The control of nature

1 These remarks raise important questions about who are the agents that exercise control and the social relations under which they do so, and the extent to which these relations include relations of control among human beings, as well as material objects (Lacey 1990; Leiss 1972).

2 Elsewhere I compared it to the relationship that Weber spoke of as an "elective infinity" between Protestantism and capitalism (Lacey 1999a, 1999b).

3 Shiva (1988, 1993) maintains that the predominance of research conducted under the materialist strategies derives from the hegemony of capitalism, to which she holds the modern values of control to be subordinate. I discuss this in Lacey (1999b).

4 Why the modern values of control have become, and remain, ascendant is a question of major importance whose exploration lies beyond the scope of this book.

5 All unaccompanied page references are to Taylor (1982).

6 The argument presented here, though it evolved from freely interpreting Taylor's argument, does not pretend to be faithful to his in its details, terminology and objectives. (See Lacey (1986) for direct analysis and discussion of his argument.)

7 For a generalization of the argument in play here see Lacey (1997c).

8 The displacement argument does not imply that there are no losses that accompany the gains of the displacement, for example, as Taylor points out, a sense of attunement with nature.

9 When Taylor returns to his argument (Taylor 1995), he seems to limit his case to the displacement argument.

10 Unless one were committed to materialist metaphysics as Taylor is not (Taylor 1970, 1985; Lacey 1990).

11 I have borrowed this broadly neo-Kantian point from Kuhn (Chapter 7). It is not completely decisive. Some of the terms of materialist lexicons may be absolute, but nothing about scientific practices assures us that this is so. Kuhn's point is compatible with the truth of scientific realism (Sankey 1997), but the practices of science themselves seem more amenable directly to carefully crafted constructivist accounts of their object of inquiry.

12 Of course we do share the world – the natural/historical world of which we are a part, which includes a "world" of social practices and objects (Chapter 7); that is why the displacement argument is sound.

13 Taylor also elaborates his argument using the notions of "disinterested representation" (Taylor 1995: 11) and "disengaged perspective" (89–90). My argument, that they do not serve to separate adoption of the materialist strategies from the link with the modern values of control, will be quite familiar by now.

7 Kuhn: scientific activity in different 'worlds'

1 See Chapter 1, note 9, on the relationship of my "strategy" to Kuhn's "paradigm." Throughout this chapter, since my concern is with Kuhn-originated ideas and not with Kuhnian scholarship, I will refer to many ideas *expressed in my terminology* as Kuhn's.

2 For the logic in play here and its limitations see MacIntyre (1988, last few chapters).

3 The developments and changes of Kuhn's position, and the wide range of interpretations of his views, are discussed critically in Hoyningen-Huene (1993) and Sankey (1994).

4 While the transition of the sixteenth and seventeenth centuries, "the scientific revolution," has been the most dramatic transformation of the kind that Kuhn discusses, he thinks that less dramatic instances are relatively common; since,

for Kuhn, there are as many scientific 'worlds' – or 'sub-worlds' – as there are fields of inquiry.

5 Sometimes A will refer to Aristotelian-type theories, sometimes to Aristotelian scientists; similarly G. In context no ambiguity will arise.

6 The notions of "acceptance," "significance" and "application" of a theory were introduced in Chapter 1.

7 No doubt it also derives (in part) from the character of general human perceptual mechanisms.

8 Conversely, we have lost much of the observational skill of A, and of numerous "pre-modern" forms of local knowledge and the practical skills associated with them.

9 It is part of the meanings of teleology and lawfulness (accord with differential equations) that they exclude each other. In G's explanations, what happens when a process is under way is explained by the initial conditions of the process (and the boundary conditions of the space) and the laws, not by a tendency to reach a final state.

10 See the "displacement" argument in Chapter 6.

11 The specifics of the planetary theories of A and G that I cite are all well documented in standard histories of the subject (e.g., Kuhn 1957).

12 A did not reject his general theory as a theory of cosmologically significant phenomena when he accepted Ptolemaic astronomy in place of Eudoxan. In this case, other things were not equal: there could be an instrumentalist accommodation, because Ptolemaic theory was not part of a more comprehensive theory that was incompatible with A with respect to a range of subdomains, it was only more empirically adequate than Eudoxan theory with respect to planetary movements and appearances, and Ptolemaic theory did not manifest the other cognitive values very highly with respect to these phenomena.

13 A queried *any* move from artifact (including experiment) to nature. The unification, brought about by Newton, of certain experimental and natural phenomena served to undermine A's query, and to legitimate the ready moves made from experiment to nature in modern science. While A's query cannot be sustained generally, it is always potentially relevant. It is always appropriate to ask whether the natural space is sufficiently like the experimental one for extrapolation to be made without further investigation of the natural space (Lacey 1984; Schwartz and Lacey 1982: Chapters 2, 9).

14 Later in the scientific tradition, especially when technological application becomes of major importance, a scientific 'world' and a social "world" might mutually support each other's development.

15 John Clendinnen (in discussion) has challenged that, with remarks like this, I mischaracterize the nature of modern (Galilean) science. See "objections" to my account of full understanding (Chapter 5).

8 A "grassroots empowerment" approach

1 In place of the term "authentic," which I borrow from Latin American discussions, others have used "alternative," "appropriate," "integral," and "popular" (Lacey 1994). ("Sustainable development" is ambiguous and open to be co-opted by versions of both modernizing and authentic development: several articles in Sachs 1993). No one of these terms has gained general currency and all of them have disadvantages (Fabián 1991; Escobar 1995;

Goulet 1988). This section represents my attempt to offer a summary synthesis of thinking that is characteristic of the popular movement in Latin America (see p. 185) on development and its processes. In part, I draw from insights taken from the "anti-development" (e.g., Escobar 1995), "liberation theology" (e.g., the essays of Ellacuría, in Hassett and Lacey 1991) and "new social movement" (e.g., Escobar and Alvarez 1992) literatures.

2 They provide the model for my fifth path toward equilibrium, "transformation from below" (Chapter 2), and they often are called "new social movements" in sociological writings.

3 The language of "human rights" – the full list: social/economic/cultural as well as civil/political – is widely used in the articulations of authentic development, often under an innovative interpretation. Human rights can be taken to represent what human beings can become; they constitute a check-point on the legitimacy of institutions that shape people's lives. Looked at this way, human rights represent aspirations, whose realizability is asserted in the modern array of rights' documents; appropriate institutions are being sought for their protection and provision, and for their absence current institutions can be held accountable (Lacey 1991b; several articles in Hassett and Lacey 1991; Faria 1994).

4 This does not imply that theory has no role in informing the struggle, especially concerning the structural obstacles to its progress and how the interaction of human nature with historical conditions shapes the range of possibilities open to serious exploration (Lacey 1997c).

5 My usage of "appropriate technology" may be narrower than customary ones (Fisher 1987).

6 This subsection summarizes parts of Shiva (1991). For documentation and further details see the writings of Shiva listed in the Bibliography; also Tiles and Oberdiek (1995: Chapter 6).

7 See also Shiva (1997) and, for some critical remarks, Lacey (1999b). Note also that, in addition to the social and ecological critique of the green revolution, a methodological critique of its research practices can also be mounted (Lacey 1998).

8 Cf. Chomsky (1993: 114–17) on the connection between imperialism and the destruction of forms of knowledge (also Harding 1998).

9 *Implementing* the green revolution itself, as distinct from its *legitimation* in the face of criticisms, does not presuppose "no lost possibilities"; alternative proposals presuppose that there are lost possibilities. The former presupposes only that the possibilities uncovered by green revolution research can be realized in practice.

Of course something is lost: local traditions are undermined and community values can no longer be aspired to. However – the objection maintains – the cause of this is not the implementations of the green revolution (and their social requirements), but the inability of traditional forms and their (possible) contemporary transformations to meet the basic needs of the poor.

10 For example, Shiva (1991), several articles in Suárez (1990) and in Delgado *et al.* (1990), Ambrose (1983), Kloppenburg (1988), Rist and San Martin (1991), Delgado (1992).

11 See especially Shiva (1988, 1991); also Levins and Lewontin (1985), Rist (1992), Brusch and Stabinsky (1996); several articles in Delgado *et al.* (1990), Marglin and Marglin (1990) and Nandy (1988).

12 Quotes from Shiva (1991: 26, 29, 97 respectively).

13 Cf. the interactionist account of the causation of tuberculosis in Lewontin (1993).

14 For a discussion of the ethical issues raised here, see Lacey (1998).

9 A feminist approach

1 I also draw upon Longino (1987, 1992, 1993, 1995, 1996), Anderson (1995a, 1995b), and discussions with Lynn Hankinson Nelson.

2 Quotes from Longino 1990: 189, 150, 171, 190 respectively.

3 Cf. the "paths to equilibrium" (Chapter 2), especially the fourth ("the quest for power") and fifth ("transformation from below"), and the highlighting of enhancing grassroots agency among the values of authentic development (Chapter 8).

4 Instead, at best they asked: "What material benefits flow from successfully adopting the stance of control?"

5 Cf. the complementary paths towards equilibrium, "adjustment" and "resignation" (Chapter 2).

6 Longino's (and Anderson's) version of feminism is very inclusive, and it can be endorsed from a wide variety of value complexes including those linked with authentic development. (If "agency" is complemented with "solidarity," a rich synthesis of O_2 and O_3 can be developed.) I take this kind of inclusiveness to be a virtue. The label "feminist" is appropriate because the values drawn upon, though defensible generally as "liberatory," have been particularly highlighted in recent feminist discussions. Furthermore, although O_3 concerns in principle all the conditions that diminish agency, its developments to date have put special (though not exclusive: e.g., Harding 1998) emphasis on those that diminish women's agency.

7 Where I use "strategy," Longino uses "global assumption." The different uses may reflect more than terminological preferences.

8 Anderson (in discussion) questions that the intentionality of action is inconsistent with its lawfulness by citing rational choice theory in economics. This is an important question and deserves more detailed attention than I can provide here. Briefly, my response is that rational choice theory remains largely an ideal type, explaining little of the interesting detail of action; and that, where it does succeed in representing action (behavior) lawfully it is in contexts in which I describe the agency as diminished, in which (for example) structural conditions constrain the available options so narrowly that the agent's adopted values are not able to be manifested to any significant degree. Then the intentional terms "preference," "belief," etc. could readily be replaced by explicitly non-intentional terms (as "purpose" is replaced by "reinforcement" in many behaviorist theories) without loss of explanatory and predictive power. On the other hand, I agree with Anderson that action, insofar as it is intentional, does have law-like features. That diminished agency can be interpreted as socially produced reflects the fact that agency has structural enabling conditions. It also has physiological enabling conditions. Physiological and social structures provide both enabling conditions and potential constraints on agency – so that in some respects action is necessarily implicated in law-like regularities.

9 Longino (1990: Chapters 6, 7). I report Longino's discussion of LHM for purposes of illustration only. (I make no effort to evaluate it.)

10 Haark (1996) seems to interpret Longino in this way (cf. Anderson 1995b).

11 Reference to and quotes from Longino 1990: 143–61, 185, 187 respectively.

12 To avoid repetition of what by now will be familiar themes I only sketch an argument for these two points.

13 Addressing arguments that intentionality is characteristic of only one stance (adopted for pragmatic ends) among others (design and physicalist) that might be adopted towards human beings (e.g., Dennett 1987) lies beyond the scope of this work.

14 From the perspective of those following O_3, this consequence is, of course, highly unlikely; and, if sufficiently general in range of application, also paradoxical (Donagan 1987; Lacey 1996).

15 As, for example, with theories of radical behaviorism and cognitive psychology.

16 For an instance of the relevant kind of argument see Schwartz, Schuldenfrei and Lacey (1978) and Schwartz and Lacey (1982).

17 If action based upon such posits is successful, that will provide further confirming evidence of the posits – just as successful technological applications provide further evidence supporting the theories that inform them. Moreover, without action committed to implementing some of the relevant possibilities (which may require considerable social change) it is impossible to explore them, for there will be no empirical basis upon which to base hypotheses about them. It is in this sense that appraisals of theories developed within the feminist approach need to pay heed to the outcomes of *political* action.

18 Such investigation might lead to qualified support for feminist "standpoint epistemologies," discussed (for example) in Harding (1996).

19 I will not discuss the epistemological framework in which Longino presents her argument.

10 Science as value free: revised theses

1 For example, the criticism of the "linear hormonal model" in Chapter 9; see also Chapter 7.

2 The account here generalizes an argument presented in Chapter 9; it is influenced by MacIntyre (1977).

3 For example, "no lost possibilities," Chapter 8.

4 The conclusions drawn about Rudner's condition (Chapter 4) apply unchanged in connection with *I*.

5 It is a value, not necessarily an over-riding one. In context, the urgencies of the moment or the necessity to act may make it subordinate to other social values, and the costs of undertaking the relevant inquiry may not warrant losses that might be incurred in other valued activities. Where it is subordinated, I suggest, the tradition of modern science has always articulated (but not always acted upon): clearly separate accepted theories from ideology; do not subordinate understanding to the interests of power – so that if one "must" act in a context where one's action is informed by, and/or its legitimation presupposes, less than soundly accepted theories (posits), be aware that this is so and monitor outcomes carefully for possible undesirable, unintended side-effects.

6 This is one reason why it is difficult to reach accord with *I* in psychology and the human sciences in general (Lacey 1980, 1997c).

7 Brian Ellis pointed out to me (in discussion) that sometimes the cognitive values operate prior to any considerations of strategies; that their play alone is sufficient to exclude from consideration certain hypotheses, e.g., that the universe came into being five minutes ago. (See also Ellis 1990.)

8 The famous cases of Lysenko and the persecution of Galileo have come to symbolize what happens when theories are accepted in violation of *I*; they turned into tragedies when power was also used to uphold these theories.

9 Examples may be found in Lewontin 1993; Longino 1990; Nelson and Nelson 1995.

10 Elsewhere, borrowing from Bhaskar (1975), I have referred to them as the conditions that define "closed spaces," spaces that in virtue of their structures ensure that, to a reasonable approximation, only a small number of specified forces and influences are acting (Lacey 1984, 1990; Lacey and Schwartz 1986, 1987). Following criticism from Marcos Barbosa de Oliveira, I now think this is too narrow a specification.

11 The mechanism involved here – call it "the overgeneralization of meta-inductions" – is a common partial source of the *de facto* violation of *impartiality*, for example, in the examples discussed by Longino (1990) and Nelson and Nelson (1995), especially those about alleged gender and racial differences in human abilities and capacities.

12 The statements of N''(1), (2) are different from those of N'(1), (2) only in minor matters of wording, abbreviation and adaptation to the context of *I*. I will leave open whether the addendum, included in (3') in the text on p. 237, should be included definitively.

13 With it, it should be kept in mind, "no lost possibilities" might be established, and then these value complexes would be rendered unviable. It is an odd notion of "neutrality" that denies potentially significant options "an (empirical) run for the money," and which leaves it for the significance to "trickle down" to where the modern values of control are not highly manifested. As in economic matters, "trickle down" is a pale shadow of "evenhandedness."

14 I take "*viable* value complex" to be redefined here as "value complex whose presuppositions are consistent with all theories accepted in accordance with *I*."

15 Accepting posits like these, without accord with *I*, and acting on them is to court disaster (especially for the poor) and exacerbation of the contemporary social and ecological crises (cf. Chapter 8).

16 My argument for endorsing N supports the exploration of multiple strategies, but – since it is based on considerations of human flourishing – it does not support the moral legitimacy of research conducted under any strategies whatsoever.

17 I will not subject item (1) (shared by N and N'') to scrutiny in this book. Recall that the materialist strategies ensure that it is true for theories developed under them by deploying lexicons that contain no value terms. That argument no longer obtains when multiple strategies are in play, especially in the social sciences whose descriptive, explanatory and anticipatory lexicons may include value-laden terms. Elsewhere (Lacey 1997c) I explore some arguments against (1) in detail.

18 In the case of N, I would urge, the social value involved makes a compelling claim on all of us – but the argument for that cannot be developed here.

19 It is not important that every individual investigator possess all these virtues, but that there be a healthy spread of them throughout the scientific community.

Bibliography

Ambrose, R. (1983) "Agricultural research and breaking the cycle of dependency," in *Science for the People* 15(6): 6–31.

Anderson, E.S. (1995a) "Feminist epistemology: an interpretation and defense," in *Hypatia* 10: 50–84.

—— (1995b) "Knowledge, human interests, and objectivity in feminist epistemology," in *Philosophical Topics* 23: 27–58.

—— (1995a) "Feminist epistemology: an interpretation and defense," in *Hypatia* 10: 50–84.

—— (1993) *Values in Ethics and Economics*, Cambridge, MA: Harvard University Press.

Bacon, F. (1620/1960) *The New Organon and Related Writings*, Indianapolis: Bobbs-Merrill.

Bernstein, R.J. (1983) *Beyond Objectivism and Relativism: science, hermeneutics and praxis*, Philadelphia: University of Pennsylvania Press.

Bhaskar, R. (1986) *Scientific Realism and Human Emancipation*, London: Verso.

—— (1979) *The Possibility of Naturalism: a philosophical critique of the contemporary human sciences*, Atlantic Highlands, NJ: Humanities Press.

—— (1975) *A Realist Theory of Science*, Atlantic Highlands, NJ: Humanities Press.

Bronowski, J. (1961) *Science and Human Values*, London: Hutchinson.

Brush, S.B. (1989) "Prediction and theory evaluation: the case of light bending," in *Science* 246: 1126–9.

Brush, S.B. and Stabinsky, D. (1996) *Valuing Local Knowledge: indigenous people and intellectual property rights*, Washington: Island Press.

Bunge, M. (1980) *Ciencia y Desarrollo*, Buenos Aires: Siglo Veinte.

Campbell, N.R. (1957) *Foundations of Science*, New York: Dover.

Carnap, R. (1967) *The Logical Structure of the World and Pseudoproblems in Philosophy*, Berkeley: University of California Press.

—— (1956) "Empiricism, semantics and ontology," in R. Carnap *Meaning and Necessity*, Chicago: University of Chicago Press.

Chomsky, N. (1993) *Year 501: the conquest continues*, Boston: South End Press.

Coady, C.A.J. (1992) *Testimony: a philosophical study*, Oxford: Clarendon Press.

Cupani, A. (1998) "A propósito do 'ethos' da ciência," in *Episteme* (Brazil) 3, 16-38.

da Costa, N. and Bueno, O. (1998) "Paraconsistência: esboço de uma interpretação," in N. da Costa, J.-Y. Béziau, and O. Bueno *Elementos de Teoria Paraconsistentes de Conjuntos*, Campinas (Brazil): Coleção CLE.

Delgado, F. (1992) *La Agroecología en las Estrategias del Desarrollo Rural*, Cusco: Centro de estudios regionales andines "Bartolomé de las Casas".

—— et al. (1990) *Agroecología y Saber Andino*, Cochabamba: AGRUCO-PRATEC.

Dennett, D.C. (1987) *The Intentional Stance*, Cambridge, MA: MIT Press.

Donagan, A. (1987) *Choice: the essential element in human action*, London: Routledge & Kegan Paul.

Dupré, J. (1993) *The Disorder of Things: metaphysical foundations of the disunity of science*, Cambridge, MA: Harvard University Press.

Eldridge, R. (1997) *Leading a Human Life: Wittgenstein, intentionality and romanticism*, Chicago: University of Chicago Press.

Ellis, B.D. (1990) *Truth and Objectivity*, Oxford: Basil Blackwell.

—— (1985) "What science aims to do," in P.M. Churchland and C.A. Hooker (eds) *Images of Science*, Chicago: University of Chicago Press.

Escobar, A. (1995) *Encountering Development: the making and unmaking of the Third World*, Princeton: Princeton University Press.

Escobar, A. and Alvarez, S.E. (eds) (1992) *The Making of Social Movements in Latin America: identity, strategy and democracy*, Boulder: Westview Press.

Fabián, R.R. (1991) "Comentários críticos a las teorias de desarrollo predominantes: hacia otro concepto del desarrollo de las comunidades," *Realidad* 4: 713–61.

Faria, J.E. (ed.) (1994) *Direitos Humanos, Direitos Socias e Justiça*, São Paulo: Malheiros Editores.

Feigl, H. (1961) "Philosophical tangents of science," in H. Feigl and G. Maxwell (eds) *Current Issues in the Philosophy of Science*, New York: Holt, Rinehart and Winston.

Feyerabend, P. (1989) "Realism and the historicity of knowledge," in *The Journal of Philosophy* 86: 393–406.

—— (1979) *Science in a Free Society*, London: New Left Books.

—— (1975) *Against Method*, London: New Left Books.

—— (1965) "Problems of empiricism," in R.S. Colodny (ed.) *Beyond the Edge of Certainty*, Englewood Cliffs, NJ: Prentice Hall.

Fisher, F.G. (1987) "Ways of knowing and the ecology of change: understanding non-technical barriers to appropriate technology," UNESCO Network for Appropriate Technology Seminar, The University of Melbourne.

Galilei, Galileo (1623/1957) *The Assayer*, excerpts in S. Drake (trans.) *Discoveries and Opinions of Galileo*, Darden City, NJ: Doubleday.

Goulet, D. (1988) "Tasks and methods in development ethics," in *Cross Currents* 38: 146–53.

Haark, S. (1996) "Science as social? – yes and no," in Nelson and Nelson (1996).

Hacking, I. (1993) "Working in a new world: the taxonomic solution," in P. Horwich (ed.) *World Changes: Thomas Kuhn and the nature of science*, Cambridge, MA: MIT Press.

—— (1983) *Representing and Intervening*, Cambridge: Cambridge University Press.

Harding, S. (1998) *Is Science Multicultural? postcolonials, feminisms, and epistemologies*, Bloomington, IN: Indiana University Press.

—— (1996) "Rethinking standpoint epistemology: what is 'strong objectivity'?" in E.F. Keller and H.E. Longino (eds) *Feminism and Science*, Oxford: Oxford University Press.

—— (ed.) (1993) *The "Racial" Economy of Science: toward a democratic future*, Bloomington, IN: Indiana University Press.

Hassett, J. and Lacey, H. (eds) (1991) *Towards a Society that Serves Its People: the intellectual contribution of El Salvador's murdered Jesuits*, Georgetown: Georgetown University Press.

Hempel, C.G. (1983a) "Values and objectivity in science," in R.S. Cohen and L. Laudan (eds) *Physics, Philosophy and Psychoanalysis*, Dordrecht: Reidel.

—— (1983b) "Kuhn and Salmon on rationality and theory choice," in *The Journal of Philosophy* 80: 570–2.

—— (1981) "Turns in the evolution of the problem of induction," in *Syntheses* 46: 389–404.

—— (1965) "Science and human values," in Hempel *Aspects of Scientific Explanation*, New York: The Free Press.

Hesse, M. (1977) "Theory and value in the social sciences," in C. Hookway and P. Pettit (eds) *Action and Interpretation*, Cambridge: Cambridge University Press.

Hoyningen-Huene, P. (1993) *Reconstructing Scientific Revolutions: Thomas S. Kuhn's philosophy of science*, Chicago: University of Chicago Press.

Hull, D.L. (1988) *Science as Process: an evolutionary account of the social and conceptual development of science*, Chicago: University of Chicago Press.

Hume, D. (1739/1968) *A Treatise of Human Nature*, Oxford: Clarendon Press.

Joseph, G. (1980) "The many sciences and the one world," in *The Journal of Philosophy* 77: 773–91.

Keller, E.F. (1982) "Feminism and science," in *Signs: Journal of Women in Culture and Society* 7: 589–602.

Kenney, M. (1986) *Biotechnology: the university – industrial complex*, New Haven: Yale University Press.

Kitcher, P. (1993) *The Advancement of Science: science without legend, objectivity without illusions*, New York: Oxford University Press.

Kloppenburg, J.R. (1988) *First the Seed: the political economy of plant biotechnology, 1492–2000*, Cambridge, Cambridge University Press.

Koyré, A. (1957) *From the Closed World to the Infinite Universe*, New York: Harper Torchbooks.

Kuhn, T.S. (1993) "Afterwords" in P. Horwich (ed.) *World Changes: Thomas Kuhn and the nature of science*, Cambridge, MA: MIT Press.

—— (1977) "Objectivity, value judgment and theory choice," in Kuhn *The Essential Tension*, Chicago: University of Chicago Press.

—— (1970) *The Structure of Scientific Revolutions*, 2nd edn, Chicago: University of Chicago Press.

—— (1957) *The Copernican Revolution: planetary astronomy in the development of Western thought*, Cambridge: Harvard University Press.

Lacey, H. (1999a) "Scientific understanding and the control of nature," in *Science and Education* 8: (13–35).

—— (1999b) "On cognitive and social values: a reply to my critics," in *Science and Education* 8: (89–103).

—— (1998) "The dialectic of science and advanced technology: an alternative?" in *Democracy and Nature* 4: 34–53.

—— (1997a) "Ciência e valores," in *Manuscrito* 20: 9–36.

—— (1997b) "The constitutive values of science," in *Princípia* 1: 3–40.

—— (1997c) "Neutrality in the social sciences: on Bhaskar's argument for an essential emancipatory impulse in the social sciences," in *Journal for the Theory of Social Behavior* 27: 213–41.

—— (1996) "Behaviorisms: theoretical and teleological," in *Behavior and Philosophy* 23: 61–78.

—— (1995) "The legacy of El Salvador's murdered Jesuits," in *Journal for Peace and Justice Studies* 6: 113–26.

—— (1994) "¿Qué tipo de ciencia le sirve al desarrollo auténtico?" *Realidad* 39: 369–82.

—— (1991a) "Understanding conflicts between North and South," in M. Dascal (ed.) *Cultural Relativism and Philosophy: North and Latin American perspectives*, Leiden: E.J. Brill.

—— (1991b) "Liberation theology and human rights," in *Proceeding from the 38th Annual Meeting of the Rocky Mountains Council of Latin American Studies*, 101–10.

—— (1990) "Interpretation and theory in the natural and human sciences," in *Journal for the Theory of Social Behavior* 20: 197–212.

—— (1986) "The rationality of science," in J. Margolis, M. Krausz and R.A. Burian (eds) *Rationality, Relativism and the Human Sciences*, Dordrecht: Kluwer Academic Press.

—— (1985) "On liberation," in *Cross Currents* 35: 219–41.

—— (1984) "Constraints upon acceptable theories: Chomsky's analogy between language acquisition and theory formation," in *Pacific Philosophical Quarterly* 65: 16–78.

—— (1980) "Psychological conflict and human nature: the case of behaviorism and cognition," in *Journal for the Theory of Social Behavior* 10: 131–55.

—— (1979) "Skinner on the prediction and control of behavior," *Theory and Decision* 10: 353–85.

—— (1974) "The scientific study of linguistic behavior: a perspective on the Skinner–Chomsky controversy," in *Journal for the Theory of Social Behavior* 4: 17–51.

Lacey, H. and Schwartz, B. (1996) "The formation and transformation of values," in W. O'Donohue and R. Kitchener (eds) *The Philosophy of Psychology*, London: Sage.

—— (1987) "The explanatory power of radical behaviorism," in S. Modgil and C. Modgil (eds) *B.F. Skinner: Consensus and controversy*, London: Falmer Press.

—— (1986) "Behaviorism, intentionality and socio-historical structure" in *Behaviorism* 14: 193–210.

Lakatos, I. (1970) "Falsification and the methodology of scientific research programmes," in I. Lakatos and A. Musgrave (eds) *Criticisms and the Growth of Knowledge*, Cambridge: Cambridge University Press.

Laudan, L. (1984) *Science and Values: the aims of science and their role in scientific debate*, Berkeley: University of California Press.

Leiss, W. (1972) *The Domination of Nature*, Boston: Beacon Press.

Levins, R. and Lewontin, R. (1985) *The Dialectical Biologist*, Cambridge: Harvard University Press.

Lewontin, R. (1993) *Biology as Ideology*, New York: Harper Perennial.

Lloyd, E.A. (1996) "Science and anti-science," in L.H. Nelson and J. Nelson (eds) (1996) *Feminism, Science and the Philosophy of Science*, Dordrecht: Kluwer Academic Press.

Longino, H.E. (1996) "Cognitive and non-cognitive values in science: rethinking the dichotomy," in L.H. Nelson and J. Nelson (eds) *Feminism, Science and the Philosophy of Science*, Dordrecht: Kluwer Academic Press.

—— (1995) "Gender, politics, and the theoretical virtues," in *Synthese* 104: 383–97.

—— (1993) "Subjects, power and knowledge: description and prescription in feminist philosophies of science," in L. Alcoff and E. Potter (eds) *Feminist epistemologies*, New York: Routledge.

—— (1992) "Essential tensions – phase two: feminist, philosophical, and social studies of science," in E. McMullin (ed.) *The Social Dimensions of Science*, Notre Dame: University of Notre Dame Press.

—— (1990) *Science as Social Knowledge*, Princeton: Princeton University Press.

—— (1987) "Can there be a feminist science?" in *Hypatia* 2: 51–64.

MacIntyre, A. (1988) *Whose Justice? Whose Rationality?* Notre Dame: University of Notre Dame Press.

—— (1981) *After Virtue*, Notre Dame: University of Notre Dame Press.

—— (1977) "Epistemological crises, dramatic narrative and the philosophy of science," in *The Monist* 60: 453–72.

Marglin, F.A. and Marglin, S.A. (eds) (1990) *Dominating Knowledge: development, culture and resistance*, Oxford: Clarinden Press.

Margolis, J. (1995) *Historied Thought, Constructed World: a conceptual primer for the turn of the millennium*, Berkeley: University of California Press.

Maxwell, N. (1984) *From Knowledge to Wisdom: a revolution in the aims and methods of science*, Oxford: Blackwell.

McDowell, J. (1994) *Mind and World*, Cambridge, MA: Harvard University Press.

McMullin, E. (1999) "Materialist categories?" in *Science and Education* 8: (37–44).

—— (1998) "A case for scientific realism," in J.A. Kourany (ed.) *Scientific Knowledge: basic issues in the philosophy of science*, 2nd edn, Belmont, CA: Wadsworth.

—— (1996) "Values in science," in W. Newton-Smith (ed.) *A Companion to the Philosophy of Science*, Cambridge: Blackwell.

—— (1993) "Rationality and paradigm change in science," in P. Horwich (ed.) *World Changes: Thomas Kuhn and the nature of science*, Cambridge, MA: MIT Press.

—— (1983) "Values in science," in P.D. Asquith and T. Nickles (eds) *PSA 1982*, vol. 2, East Lansing, MI: Philosophy of Science Association.

Merton, R. (1957) *Social Theory and Social Structure*, Glencoe: Free Press.

Midgley, M. (1979) *Beast and Man: the roots of human nature*, London: Routledge.

Murdoch, I. (1992) *Metaphysics as a Guide to Morals*, London: Penguin.

Nagel, E. (1961) *The Structure of Science*, New York: Harcourt, Brace and World.

Nandy, A. (ed.) (1988) *Science, Hegemony and Violence*, Delhi: Oxford University Press.

Nelson, L.H. (1996) "Empiricism without dogmas," in L.H. Nelson and J. Nelson (eds) *Feminism, Science and the Philosophy of Science*, Dordrecht: Kluwer Academic Press.

—— (1995) "A feminist naturalized philosophy of science" in *Synthese* 104: 399–421.

Nelson, L.H. and Nelson, J. (1995) "Feminist values and cognitive virtues," in D. Hull, M. Forbes and R.M. Burian (eds) *PSA 1994: proceedings of the 1994 biennial meeting of the Philosophy of Science Association*, vol. 2, East Lansing, MI: Philosophy of Science Association.

Nerlich, G. (1989) *Values and Valuing: speculations on the ethical life of persons*, Oxford: Oxford University Press.

Newton-Smith, W.H. (1981) *The Rationality of Science*, London: Routledge & Kegan Paul.

Poincaré, H. (1920/1958) *The Value of Science*, New York: Dover.

Popper, K.R. (1959) *The Logic of Scientific Discovery*, New York: Harper.

—— (1972) *Objective Knowledge*, London: Oxford University Press.

Putnam, H. (1990) *Realism with a Human Face*, Cambridge, MA: Harvard University Press.

—— (1987) *The Many Faces of Realism*, La Salle, IL: Open Court.

—— (1981) *Reason, Truth and History*, Cambridge: Cambridge University Press.

—— (1978) *Meaning and the Moral Sciences*, London: Routledge & Kegan Paul.

Rescher, N. (1965) "The ethical dimension of scientific research," in R.G. Colodny (ed.) *Beyond the Edge of Certainty*, Englewood Cliffs, NJ: Prentice-Hall.

Rist, S. (1992) *Desarrollo y Participacion: experiencias con la revalorizacion del conocimiento campesino en Bolivia*, Cochabamba: AGRUCO.

Rist, S. and San Martin, J. (1991) *Agroecologia y Saber Campesino en la Conservación de Suelos*, Cochabamba: AGRUCO.

Root, M. (1993) *Philosophy of Social Science*, Oxford: Blackwell.

Rouse, J. (1987) *Knowledge and Power: towards a political philosophy of science*, Ithaca: Cornell University Press.

Rudner, R. (1953) "The scientist *qua* scientist makes value judgments," in *Philosophy of Science* 20: 1–6.

Sachs, W. (1993) *Global Ecology: a new arena of political conflict*, London: Zed Books.

Salmon, W.C. (1983) "Carl G. Hempel on the rationality of science," in *The Journal of Philosophy* 80: 555–62.

—— (1966) *The Foundations of Scientific Inference*, Pittsburgh: University of Pittsburgh Press.

Sankey, H. (1997) *Rationality, Relativism and Incommensurability*, Aldershot: Avebury.

—— (1994) *The Incommensurability Thesis*, Aldershot: Avebury

Schwartz, B. (1997) "Psychology, idea technology, and ideology," in *Psychological Science* 8: 21–7.

—— (1986) *The Battle for Human Nature: science, morality and modern life*, New York: Norton.

Schwartz, B. and Lacey, H. (1982) *Behaviorism, Science and Human Nature*, New York: Norton.

Schwartz, B., Schuldenfrei, R. and Lacey, H. (1978) "Operant psychology as factory psychology," in *Behaviorism* 6: 29–54.

Scriven, M. (1991) *Evaluation Thesaurus*, 4th edn, Newbury Park, CA: Sage.

—— (1974) "The exact role of value judgments in science," in R.S. Cohen and K. Schaffner (eds) *Proceedings of the 1972 Biennial Meeting of the Philosophy of Science Association*, Dordrecht: Reidel.

Shiva, V. (1997) *Biopiracy: the plunder of nature and knowledge*, Boston: South End Press.

—— (1993) *Monocultures of the Mind: perspectives on biodiversity and biotechnology*, London: Zed.

—— (1991) *The Violence of the Green Revolution*, London: Zed.

—— (1989) *Staying Alive: women, ecology and development*, London: Zed.

—— (1988) "Reductionist science as epistemological violence," in A. Nandy (ed.) *Science, Hegemony and Violence*, Delhi: Oxford University Press.

Skinner, B.F. (1972) "Are theories of learning necessary?" in Skinner *Cumulative Record*, 3rd edn, New York: Appleton-Century-Crofts.

Solomon, M. (1994) "Social empiricism" in *Noûs* 28: 325–43.

—— (1992) "Scientific rationality and human reasoning," in *Philosophy of Science* 59: 439–55,

Suárez, B. (ed.) (1990) *¿Biotechnologia para el Progreso Agricola de México?*, Mexico City: Centro de Ecodesarrollo.

Taylor, C. (1995) *Philosophical Arguments*, Cambridge: Harvard University Press.

—— (1985) *Human Agency and Language: philosophical papers*, vol. 1, Cambridge: Cambridge University Press.

—— (1982) "Rationality," in M. Hollis and S. Lukes (eds) *Rationality and Relativism*, Cambridge, MA: MIT Press.

—— (1981) "Understanding in the *Geisteswissenschaften*," in S.H. Holtzmann and C.M. Leich (eds) *Wittgenstein: To follow a rule*, Cambridge: Cambridge University Press.

—— (1980) "Understanding in human science," in *The Review of Metaphysics* 34: 24–38; 47–55.

—— (1971) "Interpretations and the sciences of man," in *The Review of Metaphysics* 25: 3–51.

—— (1970) "The explanation of purposive behavior," in R. Berger and F. Cioffi (eds) *Explanation in the behavioral sciences*, Cambridge: Cambridge University Press.

Tiles, M. (1987) "A science of Mars or Venus" in *Philosophy* 62: 293–306.

—— (1986) "*Mathesis* and the masculine birth of time," in *International Studies in the Philosophy of Science* 1: 16–35.

—— (1985) "Correcting concepts," in *Ratio* 27: 19–35.

Tiles, M. and Oberdiek, H. (1995) *Living in a Technological Culture: human tools and human values*, London: Routledge.

van Fraassen, B.C. (1980) *The Scientific Image*, Oxford: Clarendon Press.

Weber, M. (1949) *The Methodology of the Social Sciences*, New York: The Free Press.

White Jr, L. (1968) *Machina ex Deo*, Cambridge, MA: MIT Press.

Index